LEVELS

D1580596

Key Skills
Application of Number

Diane Parker
Sue Broadhouse
Dave Faulkner
John Gillespie

Hodder & Stoughton

A MEMBER OF THE

Orders: please contact Bookpoint Ltd, 130 Milton Park, Abingdon, Oxon OX14 4SB. Telephone: (44) 01235 827720. Fax: (44) 01235 400454. Lines are open from 9.00–6.00, Monday to Saturday, with a 24 hour message answering service.

British Library Cataloguing in Publication Data
A catalogue record for this title is available from the British Library

ISBN 0340 801484

First published 2002

Impression number 10 9 8 7 6 5 4 3 2 1
Year   2007 2006 2005 2004 2003 2002

Cover illustration by Stewart Larking

Typeset Pantek Arts Ltd
Printed in Great Britain for Hodder & Stoughton Educational, a division of Hodder Headline Plc, 338 Euston Road, London NW1 3BH by Bath Press Ltd, Bath.

Contents

Welcome to Key Skills Application of Number

This book is one of three in the *Key Skills Builder* series, and
is designed to help you understand, develop and apply your
Application of Number skills to nationally recognised
standards. Step by step, this book will help you improve your
understanding of what you need to know and do in order to
successfully achieve a Key Skills Unit Qualification in
Application of Number at either Level 1, 2 or 3.

Who needs Key Skills?

It might surprise you to learn that *everybody* needs key skills.
This is because they form an important part of work, study
and everyday life, for everyone. Although there are a total of
six key skills, the *Key Skills Builder* series concentrates on
three. These are:
- Communication
- Application of Number
- Information Technology

Each of these key skills is assessed through the production of
a portfolio and by taking a test. As a result, each of them can
be regarded as a qualification in its own right. Together they
also make up a completely new, nationally recognised
qualification called the Key Skills Qualification.

It is important to remember that there are also three other key
skills, known as the Wider Key Skills. These are:
- Working with Others
- Improving Own Learning and Performance
- Problem Solving

These are often referred to as personal skills and, unlike
the other three, are assessed only through the production
of a portfolio.

Getting the measure of Key Skills

The ability to communicate, use number and use computers
is so much a part of everyday life, that most people take these
skills completely for granted. Perhaps it is for this reason that
we rarely stop to consider these skills, or how important they
are in all our lives.

For some time, the government and employers have been
concerned that many people cannot use academic skills in

practical situations. For example, someone who has a good grade in GCSE maths may find it hard to work out a profit margin on a particular product, or someone who has passed 'A' level English finds it difficult to write formal reports for their manager at work.

In order to address these concerns, the regulatory authorities in England, Wales and Northern Ireland (QCA, ACCAC & CCEA) have published specifications for each of the key skills units, describing in detail what candidates need to know, and do, at each level. These units set national standards, which candidates can measure their performance against.

In order to demonstrate your competence in Application of Number you will need to learn more about what is involved (*check your skills*), develop the skills needed to be able use your knowledge effectively (*practise your skills*), then undertake activities that will allow you to demonstrate your competence (*apply your skills*).

It is important to remember that everybody is different, and that we all find some skills harder to apply than others. With this in mind the key skills units have been written at different levels in order to accommodate those with a basic grasp of the skill to those who can apply the skill in much more complex ways.

Whilst there are also key skills units at Levels 4 and 5, these are aimed at graduate level candidates and are therefore not covered in this particular book.

Achieving a Key Skills Qualification

Whether your target is one, two, or all three units it is important that you work towards the most suitable level for your own needs. This might well mean that you work towards different key skills at different levels (i.e. Number and Communication at Level 2 and IT at Level 3); therefore it is important that you start by discussing your plans with your tutor. Having established the most appropriate level for each key skill you need to decide/agree the most suitable contexts in which you could develop and demonstrate the skills.

For those at school or college these might include academic or vocational subjects (GCSEs AS/A2 Levels, AVCEs etc), entitlement or enrichment programmes (PSHE, citizenship, careers etc), community activities (voluntary work, youth work, charitable work, etc). Those in training or employment might well identify suitable contexts from within their working environment, whilst candidates from all backgrounds could identify suitable extra curricular and leisure activities (Duke of Edinburgh's Award, Young Enterprise, sport, music, drama productions, etc, community activities, voluntary work, youth work, charitable work, etc.) Opportunities to develop and apply key skills exist almost everywhere.

Using this book

This book contains information that will help you to decide just how much you already know and how much you still need to learn about Application of Number at each level. It describes activities that other students have used to help them develop their own skills, and includes examples of the types of evidence they produced for their portfolios. It includes checklists and action plans, and provides guidance on how to put together your portfolio and how to prepare for the tests. *Key Skills Application of Number* is designed to be used flexibly, so you can either follow the whole process through at a particular level, or dip in and out of appropriate sections as the need arises. Whether used as a structured programme of skills development, or simply as a source of information, ideas and guidance, make sure that you discuss with your tutor or supervisor how to make best use of Key Skills Application of Number.

Introduction to Application of Number

Why do we need Application of Number skills?

Application of Number skills are essential in every part of our daily lives. They are the key number applications that are found in every aspect of work or study. You will have learnt mathematical skills at school or college. You will not always have been shown how to use these skills in practical situations. Application of Number shows you how to use number skills in your studies or your place of work.

What are Application of Number Skills?

The skills required for:
- Interpreting information involving numbers
- Carrying out calculations
- Interpreting results and presenting findings

How can this book help me to develop my Application of Number Skills?

This book will help you to check that you have the knowledge and skills needed in Application of Number. It will also help you to demonstrate that you can apply these skills in different situations. You will firstly need to decide which level is best for you based on your current experience. This book will help you to do this. It will give you the best opportunity to achieve in Application of Number.

How will I be assessed in Application of Number?

As an Application of Number candidate you will be assessed in two ways before you can be given a QCA certificate:

1 **You will complete a test (this is the external assessment)**

2 **You will build a portfolio of evidence (this is the internal assessment)**

The portfolio will contain the evidence showing that you have covered what is required, and that you have reached the necessary standard. Each portfolio will be different, depending on where and how you have collected your evidence.

DEFINITIONS ▶

CANDIDATE:	Anyone registered for one or more Key Skills units. You must be registered before you can receive a certificate.
QCA:	The Qualifications and Curriculum Authority, the government organisation which has written the Key Skills units, and has decided on how they should be assessed.
ASSESSMENT:	The ways in which the skills and knowledge you have are tested and marked.
PORTFOLIO:	File of evidence which proves the things you say you can do are true.
EVIDENCE:	This could be records of calculations, interpretation of results, graphs, charts and diagrams.
LEVEL:	The Key Skill unit Application of Number can be achieved at levels 1–3.
EXTERNAL ASSESSMENT:	Test which is either multiple choice, at levels 1 and 2, or short answer questions at level 3.
INTERNAL ASSESSMENT:	The portfolio of evidence

What is the aim of the Application of Number units?

To encourage you to develop and demonstrate your skills in:
- interpreting information involving numbers
- carrying out calculations
- interpreting results and presenting your findings

They are designed to develop and recognise your ability to **select** and **apply** numerical, graphical and related mathematical skills in ways that are appropriate in your particular context. The units will also help you understand how number can be applied in areas that are less familiar to you. They can be used to develop your ability to progress to higher levels of competence.

The units cover essential techniques such as being able to:
- measure and read scales
- carry out specific calculations
- draw a particular type of diagram

Also included are the important skills of application, such as:
- interpreting information from tables, graphs or charts
- selecting appropriate methods
- describing what findings show, etc.

Techniques and skills of application both contribute to understanding a task or problem and to deciding on the best course of action.

What does the Application of Number unit look like?

Like all the Key Skills units, Application of Number can be achieved at Levels 1–3. At each level the Application of Number unit has four parts:
- **An introduction to the unit**
- **Part A – What you need to know**
- **Part B – What you must do**
- **Part C – Guidance**

An introduction to the unit

This part of the unit provides a brief summary of what the unit is about and how it is structured.

Part A – What you need to know

A description of what you need to know how to do, in order to have the confidence to apply your skills appropriately. You will need to be confident in using all the skills. You will be assessed on Part A in the test.

Part B – What you must do

An outline of what you must produce for your portfolio: how much evidence you need, and the type of evidence required. This evidence will show that you have the Application of Number skills at your chosen level. It will demonstrate your ability to apply number skills in practical situations. The criteria for portfolio assessment are included in Part B. The criteria must be used together, as a set, for each component of the unit.

Part C – Guidance

This gives you some ideas about general activities that you might use to develop and demonstrate your skills. It also provides examples of evidence that you could produce to show that you have the skills required.

What do I need to do in order to achieve an Application of Number unit?

In order to achieve the Application of Number unit you will need to:

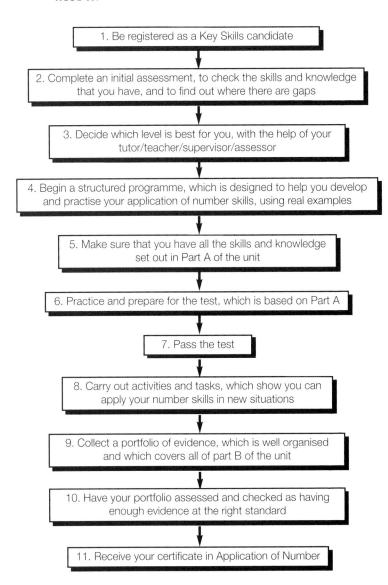

1. Be registered as a Key Skills candidate

2. Complete an initial assessment, to check the skills and knowledge that you have, and to find out where there are gaps

3. Decide which level is best for you, with the help of your tutor/teacher/supervisor/assessor

4. Begin a structured programme, which is designed to help you develop and practise your application of number skills, using real examples

5. Make sure that you have all the skills and knowledge set out in Part A of the unit

6. Practice and prepare for the test, which is based on Part A

7. Pass the test

8. Carry out activities and tasks, which show you can apply your number skills in new situations

9. Collect a portfolio of evidence, which is well organised and which covers all of part B of the unit

10. Have your portfolio assessed and checked as having enough evidence at the right standard

11. Receive your certificate in Application of Number

■ NOTE: You will be working on both Part A and Part B at the same time in your course or programme. You do not have to pass the test before your portfolio is assessed. You will, however, need to have developed and practised your skills before you take the test or put your portfolio together.

DEFINITIONS ▶

CRITERIA: The standards you must reach in the test, and in your portfolio. If you do not do well enough to meet the criteria you will not get a certificate

AWARDING BODY: The exam board with whom you are registered, which sets and marks the tests, and checks the portfolio. e.g. AQA, Edexcel, OCR, C & G, ASDAN, CCEA, CACHE,EAL, HAB, LCCI

Introduction to Levels 1–3

Application of Number unit at Levels 1, 2 and 3

You can move up through the Key Skills levels in the following ways:

1 Carry out substantial tasks, involving multi-stage calculations

2 Take more responsibility for planning, and decision making when carrying out tasks and solving problems

3 Evaluate your own performance, and understand the reasons for difficulties and successes when solving problems

What will I need to do at Level 1?

 At this level you must be able to:
- Interpret straightforward information
- Carry out calculations, using whole numbers, simple decimals, fractions and percentages to given levels of accuracy
- Interpret the results of your calculations and present findings, using a chart and diagram

You are expected to produce evidence for each component of the unit.
- All your calculations must be clearly set in context – exercises are not acceptable
- You must show that you understand why you are obtaining information and carrying out calculations
- You must describe how your results meet the purpose of the task

What will I need to do at Level 2?

 At this level you must be able to carry through a substantial activity that requires you to:
- Select information and methods to get the results you need;
- Carry out calculations involving two or more steps and numbers of any size, including use of formulae, and check your methods and levels of accuracy

- *Select ways of presenting your findings, including use of a graph, describe the methods used, and explain your results*

You are expected to carry out *at least one substantial activity that can be broken down into tasks for each component of the unit*.
- All calculations must be clearly set in context
- All calculations must be clearly set out with evidence of checking
- You must show that you are clear about the purpose of your task

What will I need to do at Level 3?

 At this level you must be able to plan and carry through a *substantial and complex activity* that requires you to:
- *Plan your approach to obtaining and using information, choose appropriate methods for obtaining the results needed, and justify your choice*
- Carry out *multi-stage calculations, including use of a large data set (over 50 items) and re-arrangement of formulae*
- *Justify your choice of presentation methods* and explain the results of your calculations

You are expected to demonstrate your skills in the context of *at least one substantial and complex activity that can be broken down into a series of interrelated tasks covering all three components of the unit.*
- Multi-stage calculations should be carried out
- All calculations must be clearly set in context
- You must show that you are clear about the purpose of each activity
- You must show that you have considered the nature and sequence of tasks when planning how to obtain and use information to meet your purpose

How will I know which Level?

 You need to find out:
- What skills you already have
- What skills you need to revise
- What skills you need to learn

How can I find out what skills I already have?

To help you check whether Level 1 is best for you, try to answer the questions, as honestly as you can, and circle the face that most reflects your feeling about the topic.

☺ Circle the happy face if you feel confident that you can demonstrate the skill easily

☺ Circle the straight face if you think you need to practise the skill

☹ Circle the sad face if you need to learn the skill

 Ask your tutor/supervisor for help if there is anything you do not understand.

Initial Self-assessment for Application of Number Level 1

Interpreting information

Circle the face

☺ **I can do this easily!**   ☺ **I need to practise**   ☹ **I need to learn**

☺ ☺ ☹   Can you read and understand straightforward tables, charts, diagrams and line graphs?

☺ ☺ ☹   Can you read and understand numbers used in different ways, (*e.g. large numbers in figures or words, simple fractions, decimals, percentages*)

☺ ☺ ☹   Can you measure in everyday units (*e.g. millimetres, minutes, litres, grams, degrees*) by reading scales on familiar equipment (*e.g. watch, tape measure, measuring jug, weighing scales, thermometer*)?

☺ ☺ ☹   Can you make accurate observations (*e.g. count number of people or items*)?

☺ ☺ ☹   Can you identify suitable calculations to get the results needed for the task?

Initial Self-assessment for Application of Number Level 1

Carrying out calculations

Circle the face

☺ **I can do this easily!**   ☺ **I need to practise**   ☹ **I need to learn**

☺ ☺ ☹   Can you work to the level of accuracy you are told to use (*e.g. round to the nearest whole unit, nearest 10, two decimal places*)?

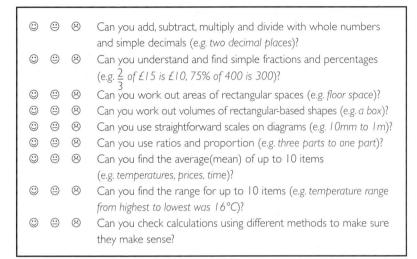

☺ ☺ ☹ Can you add, subtract, multiply and divide with whole numbers and simple decimals (e.g. *two decimal places*)?

☺ ☺ ☹ Can you understand and find simple fractions and percentages (e.g. $\frac{2}{3}$ *of £15 is £10, 75% of 400 is 300*)?

☺ ☺ ☹ Can you work out areas of rectangular spaces (e.g. *floor space*)?

☺ ☺ ☹ Can you work out volumes of rectangular-based shapes (e.g. *a box*)?

☺ ☺ ☹ Can you use straightforward scales on diagrams (e.g. *10mm to 1m*)?

☺ ☺ ☹ Can you use ratios and proportion (e.g. *three parts to one part*)?

☺ ☺ ☹ Can you find the average(mean) of up to 10 items (e.g. *temperatures, prices, time*)?

☺ ☺ ☹ Can you find the range for up to 10 items (e.g. *temperature range from highest to lowest was 16°C*)?

☺ ☺ ☹ Can you check calculations using different methods to make sure they make sense?

Initial Self-assessment for Application of Number Level 1

Interpreting results and presenting your findings

Circle the face

☺ **I can do this easily!**   ☺ **I need to practise**   ☹ **I need to learn**

☺ ☺ ☹ Can you use suitable ways of presenting information, including a chart and a diagram?

☺ ☺ ☹ Can you use the correct units (e.g. *for area, volume, weight, time, temperature*)?

☺ ☺ ☹ Can you label your work correctly (e.g. *use a title or a key*)?

☺ ☺ ☹ Can you describe how the results of your calculations meet the purpose of your task?

When you have completed the self-assessment take it to your tutor/supervisor.

What does this show?

If you have circled happy faces for most of the questions, you should be able to achieve Application of Number at Level 1. You may think that Level 1 would be too easy, and that you ought to be working at the next level, Level 2.

Check this with your tutor/supervisor

If you have circled straight faces for most of the questions, Level 1 is definitely the right level for you to work at.

 Check this with your tutor/supervisor

If you have circled sad faces for most or all of the questions, you may not be ready to try this level of Application of Number yet.

 Check this with your tutor/supervisor

What do I do next?

Draw up an action plan
Showing:
* What skills you need to learn or improve
* How you will learn or improve the skills
* The timetable within which you will do this

What do I do then?

FOLLOW YOUR ACTION PLAN!

In order to achieve Application of Number at Level 1 you will need to:

1 **Make sure you can do all the points in Part A**

2 **Practise and prepare for the test**

3 **Pass the test at Level 1**

4 **Collect a portfolio of evidence, which is well organised and which shows you can do all of Part B of the unit**

5 **Have your portfolio assessed and checked as having enough evidence at the right standard**

Summary of Application of Number Level 1

You must show that you can:
* Interpret straightforward information
* Carry out calculations, using whole numbers, simple decimals, fractions and percentages to given levels of accuracy

- Interpret the results of your calculations and present findings, using a chart and diagram

Assessment:

External test: set and marked by Awarding Body
- Test is one hour long
- Questions are multiple choice
- Up to 40 questions
- Based on Part A

Internal portfolio: assessed within your centre:
- Must cover all of Part B
- Must meet the standard
- Must be well organised

How can I find out what skills I already have at Level 2?

To help you check whether Level 2 is best for you, try to answer the questions, as honestly as you can, and circle the face that most reflects your feeling about the topic.
☺ Circle the happy face if you feel confident that you can demonstrate the skill easily
☺ Circle the straight face if you think you need to practise the skill
☹ Circle the sad face if you need to learn the skill

 Ask your tutor/supervisor for help if there is anything you do not understand.

Initial Self-assessment for Application of Number Level 2

Interpreting information
Circle the face
☺ **I can do this easily!**   ☺ **I need to practise**   ☹ **I need to learn**

☺	☺	☹	I can obtain relevant information from different sources (e.g. *from written and graphical material, first-hand by measuring or observing*)
☺	☺	☹	I can read and understand graphs, tables, charts and diagrams (e.g. *frequency diagrams*)
☺	☺	☹	I can read and understand numbers used in different ways, including negative numbers (e.g. *for losses in trading, low temperatures*)
☺	☺	☹	I can estimate amounts and proportions
☺	☺	☹	I can read scales on a range of equipment to given levels of accuracy (e.g. *to the nearest 10mm or nearest inch*)

☺ ☺ ☹ I can make accurate observations (e.g. *count the number of customers per hour*)

☺ ☺ ☹ I can select appropriate methods for obtaining the results I need, including grouping data when this is appropriate (e.g. *heights, salary bands*)

Initial Self-assessment for Application of Number Level 2

Carrying out calculations

Circle the face

☺ **I can do this easily!**   ☺ **I need to practise**   ☹ **I need to learn**

☺ ☺ ☹ I can show clearly my methods of carrying out calculations and give the level of accuracy of my results

☺ ☺ ☹ I can carry out calculations involving two or more steps, with numbers of any size

☺ ☺ ☹ I can convert between fractions, decimals and percentages

☺ ☺ ☹ I can convert measurements between systems (e.g. *from pounds to kilograms, between currencies*)

☺ ☺ ☹ I can work out areas and volumes (e.g. *area of an L-shaped room, number of containers to fill a given space*)

☺ ☺ ☹ I can work out dimensions from scale drawings (e.g. *using a 1:20 scale*)

☺ ☺ ☹ I can use proportion and calculate using ratios where appropriate

☺ ☺ ☹ I can compare sets of data with a minimum of 20 items (e.g. *using percentages, using mean, median, mode*)

☺ ☺ ☹ I can use range to describe the spread within sets of data

☺ ☺ ☹ I can understand and use given formulae (e.g. *for calculating volumes, areas such as circles, insurance premiums, V=IR for electricity*)

☺ ☺ ☹ I can check my methods in ways that pick up faults and make sure that my results make sense

---

Initial Self-assessment for Application of Number Level 2

Interpreting results and presenting your findings

Circle the face

☺ **I can do this easily!**   ☺ **I need to practise**   ☹ **I need to learn**

☺ ☺ ☹   I can select effective ways to present my findings

☺ ☺ ☹   I can construct and use graphs, charts and diagrams, (e.g. *pie charts, frequency tables, workshop drawings*) and follow accepted conventions for labelling these (e.g. *appropriate scales and axes*)

☺ ☺ ☹   I can highlight the main points of my findings and describe my methods

☺ ☺ ☹   I can explain how the results of calculations meet the purpose of my activity

When you have completed the self-assessment take it to your tutor/supervisor.

What does this show?

If you have circled happy faces for most of the questions, you should be able to achieve Application of Number at Level 2. You may think that Level 2 would be too easy, and that you ought to be working at the next level, Level 3.

! *Check this with your tutor/supervisor*

If you have circled straight faces for most of the questions, Level 2 is definitely the right level for you to work at.

! *Check this with your tutor/supervisor*

If you have circled sad faces for most or all of the questions, you may not be ready to try this level of Application of Number. You may need to start at Level 1. Try the self-assessment at Level 1 to find out.

! *Check this with your tutor/supervisor*

What do I do next?

Draw up an action plan
Showing:
● What skills you need to learn or improve
● How you will learn or improve the skills
● The timetable within which you will do this

What do I do then?

FOLLOW YOUR ACTION PLAN!

In order to achieve Application of Number at Level 2 you will need to:

1 **Make sure you can do all the points in Part A**

2 **Practise and prepare for the test**

3 **Pass the test at Level 2**

4 **Collect a portfolio of evidence, which is well organised and which shows you can do all of Part B of the unit**

5 **Have your portfolio assessed and checked as having enough evidence at the right standard**

Summary of Application of Number Level 2

You must show that you can:
● Select information and methods to get the results you need
● Carry out calculations involving two or more steps and numbers of any size, including use of formulae, and check your methods and levels of accuracy
● Select ways of presenting your findings, including use of a graph, describe methods and explain results

Assessment:

External test: set and marked by Awarding Body
● Test is one hour long
● Questions are multiple choice
● Up to 40 questions
● Based on Part A

Internal portfolio assessed within your centre:
- Must cover all of Part B
- Must meet the standard
- Must be well organised

How can I find out what skills I already have at Level 3?

To help you check whether Level 3 is best for you, try to answer the questions, as honestly as you can, and circle the face that most reflects your feeling about the topic.

☺ Circle the happy face if you feel confident that you can demonstrate the skill easily

☺ Circle the straight face if you think you need to practise the skill

☹ Circle the sad face if you need to learn the skill

 Ask your tutor/supervisor for help if there is anything you do not understand.

Initial Self-assessment for Level 3

Planning an activity and interpreting information

Circle the face

☺ **I can do this easily!**   ☺ **I need to practise**   ☹ **I need to learn**

☺	☺	☹	I can plan a substantial and complex activity by breaking down into a series of tasks
☺	☺	☹	I can obtain and use relevant information from different sources, including a large data set (over 50 items) and use this to meet the purpose of my activity
☺	☺	☹	I can use estimation to help me plan, multiplying and dividing numbers of any size (rounded to one significant figure)
☺	☺	☹	I can make accurate and reliable observations over time and use equipment to measure in a variety of appropriate units
☺	☺	☹	I can read and understand scale drawings, graphs, complex tables and charts
☺	☺	☹	Read and understand ways of writing very large and very small numbers (e.g. £1.5 billion, 2.4×10^{-3})
☺	☺	☹	I can read and use compound measures (e.g. *speed in kph pressures in psi, concentrations in ppm*)
☺	☺	☹	I can choose appropriate methods for obtaining the results I need and justify my choice

Initial Self-assessment for Level 3

Carrying out calculations

Circle the face

☺ **I can do this easily!**　　☺ **I need to practise**　　☹ **I need to learn**

☺　☺　☹　I can show my methods clearly and work to appropriate levels of accuracy

☺　☺　☹　I can carry out multi-stage calculations with numbers of any size (e.g. *find the results of growth at 8% over three years, find the volume of water in a swimming pool*)

☺　☺　☹　I can use powers and roots (e.g. *to work out interest on £5000 at 5% over three years*)

☺　☺　☹　I can work out missing angles and sides in right-angled triangles from known sides and angles

☺　☺　☹　I can work out proportional change (e.g. *add VAT at 17.5% by multiplying by 1.175*)

☺　☺ · ☹　I can work out actual measurements from scale drawings (e.g. *room or site plan, map, workshop drawing*) and scale quantities up and down

☺　☺　☹　I can work with large data sets using measures of average and range to compare distributions, and estimate mean, median and range of grouped data

☺　☺　☹　I can rearrange and use formulae, equations and expressions (e.g. *formulae in spreadsheets, finance, and area and volume calculations*)

Initial Self-assessment for Level 3

Interpreting results and presenting my findings

Circle the face

☺ **I can do this easily!**　　☺ **I need to practise**　　☹ **I need to learn**

☺　☺　☹　I can select and use appropriate methods to illustrate findings, show trends and make comparisons

☺　☺　☹　I can examine critically, and justify, my choice of methods

☺　☺　☹　I can construct and label charts, graphs, diagrams and scale drawings using accepted conventions

☺　☺　☹　I can draw appropriate conclusions based on my findings, including how possible sources of error might have affected my results

☺　☺　☹　I can explain how my results relate to the purpose of my activity

When you have completed the self-assessment take it to your tutor/supervisor.

What does this show?

If you have circled happy faces for most of the questions, you should be able to achieve Application of Number at Level 3.

Check this with your tutor/supervisor

If you have circled some happy faces but there are gaps in your skills and knowledge, this may be the most suitable level for you.

Check this with your tutor/supervisor

If you have circled sad faces for most or all of the questions, you may not be ready to try this level of Application of Number. You may need to start at Level 1, or 2. Try the self-assessment at Levels 1 and 2 to find out.

Check this with your tutor/supervisor

What do I do next?

Draw up an action plan
Showing:
- What skills you need to learn or improve
- How you will learn or improve the skills
- The timetable within which you will do this

What do I do then?

FOLLOW YOUR ACTION PLAN!

In order to achieve Application of Number at Level 3 you will need to:

1 Make sure you can do all the points in Part A

2 Practise and prepare for the test

3 Pass the test at Level 3

4 Collect a portfolio of evidence, which is well organised and which shows you can do all of Part B of the unit

5 Have your portfolio assessed and checked as having enough evidence at the right standard

Summary of Application of Number Level 3

You must show that you can:

- Plan your approach to obtaining and using information, choose appropriate methods for obtaining the results needed and justify your choice
- Carry out multi-stage calculations, including use of a large data set(over 50 items) and re-arrangement of formulae
- Justify your choice of presentation methods and explain the results of your calculations

Assessment:

External test set and marked by Awarding Body

- Test is one-and-a-half hours long
- Questions are both short answer and extended
- Based on Part A and Part B

Internal portfolio assessed within your centre:

- Must cover all of Part B
- Must meet the standard
- Must be well organised

3

Application of Number Level 1: Developing the Skills and Knowledge in Part A

Introduction

Part A describes the skills and knowledge that you need to **learn** for the Application of Number qualification at Level 1. You may have to pass a test as part of your qualification.

At Level 1, you handle straightforward numerical information. You need to know basic techniques for obtaining data, such as measuring and being able to read a graph or chart. You must be able to carry out calculations that have two or three steps. You must be able to choose the best way to give the answers and show people what the answers mean.

You develop these skills in Part A and apply them in Part B. We are going to look at Part A first.

Understanding Part A

Read through the list of "bullet points" in Part A of the specification. Probably there will be some things you think you can do, and some things you are not sure about. If you have not already done it, turn to the **self-assessment check-list** for Level 1 on page 8 and complete it. To pass the test you need to be sure that you can do all of Part A.

The test is a multiple-choice paper. It has 40 questions. The questions are designed to find out if you can do all of the things listed in Part A.

To get the right answers you have to:

• Read and understand the information in the question. Sometimes numbers are in the text of the question. Sometimes there is a table, chart, graph or diagram that you must interpret

- Do a calculation with the numbers you obtained
- Select the correct answer from the options, A, B, C or D

You are not allowed to use a calculator in this test. You can jot down the information, but then you must do the calculation in your head.

The next section helps you to check how much you already know, and what you need to practice.

As you go through this section, work out the answers to the questions **without** a calculator, then check your answers. Make sure that you understand **how** the correct answer is worked out.

Numbers used in different ways

To use all kinds of numbers you need to understand

- The base 10 or 'units, tens and hundreds' system
- The decimal point
- Units of measurement
- How to divide large units such as metres into smaller units, such as centimetres, or the reverse
- The use of symbols for area and volume
- Ways of writing money, decimal numbers, fractions, percentages

The way we write or say a number depends on what kind of information we want to give.

Here are some examples:

£5.99 is said "five, ninety nine". We are used to seeing prices given this way, so we know that it means 'five pounds and ninety nine pence.

9.40 am tells us what time it is. We say "nine forty am". On a 24-hour clock, the same time could be written as 0940, or we could give the time as "twenty to ten in the morning".

5.75 m We could read this as "five point seven five metres". Or we could give the length as five metres and seventy-five centimetres.

(Help) desk

Remember to give all the information that is needed:

- The numbers
- The decimal point
- The symbols

Read the information below, and notice how the number is written on the cheque.

A road development costs two billion pounds (two thousand million pounds).

Look again at the very large number above £2,000,000,000.

Notice how the commas divide the number into groups of three digits. Commas are not always included. Sometimes we leave small spaces. The number could be written £2 000 000 000.

Help desk

number	the whole amount
digit	part of a number, it occupies one place in the number

Try writing the numbers below as numbers, using digits.

(i)   three
(ii)   twenty two
(iii)   four thousand, nine hundred and forty eight
(iv)   five hundred thousand, six hundred and thirty
(v)   ninety million

Answers:

(i)   3
(ii)   22
(iii)   4,948   or   4 948
(iv)   500,630   or   500 630
(v)   90,000,000   or   90 000 000

Help desk

If you got any wrong, check with the table below.
Make sure you entered the first digit in the correct place.
Fill up the remaining places to the right
Enter a zero if necessary so that every place is filled

thousand millions (billions)	hundred millions	ten millions	millions	hundred thousands	ten thousands	thousands	hundreds	tens	units
2,	0	0	0,	0	0	0,	0	0	0

Decimals

The decimal point allows us to divide one unit into tenths, hundredths, thousandths, etc. These are called the **decimal places**.

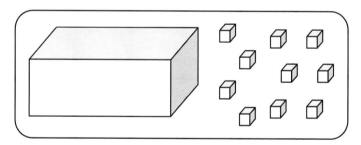

Look at this number **24.125**

> The whole number is twenty four (24 is two tens and four units)
> Then we write the decimal point
> The first decimal place is one tenth
> The second decimal place is two hundredths
> The third decimal place is five thousandths

When we say a decimal number, we just say "point" where the decimal point comes. 24.125 is said '*twenty four point one two five*'.

Try writing these numbers as decimal numbers.
(i) ten point five
(ii) three point two five

(iii) seven tenths

(iv) eight hundredths

(v) four tenths, two hundredths and five thousandths

Answers:

(i)   10.5

(ii) 3.25

(iii) 0.7

(iv) 0.08

(v) 0.425

Help desk

If you got any wrong, check back with the table above and make sure you have entered figures in the correct decimal places. Also notice two important things:

● You must have something to the left of the decimal point. If the number is less than one, enter zero in the place to the left of the decimal point.

● Always fill up the decimal places from the decimal point. Enter zero in a place where there is no other digit.

whole number		.	1st d.p.	2nd d.p.	3rd d.p.
tens	units	**decimal point**	tenths	hundredths	thousandths

Rounding numbers

We can have any number of decimal places. Sometimes the result of a calculation gives us a number that goes on for ever. Remember, the answers you give must make sense.

The more decimal places there are in a number, the greater its level of accuracy.

For example, if you measure something with a ruler, you can't really be more accurate than one millimetre (mm). It does not make sense to give a result to more decimal places than the accuracy you can get.

At Level 1, you will be told what level of accuracy to use.

Example 1

The result of a calculation comes to 3.478 Give this to 2 d.p.

The first two decimal places are .47 but we can't just forget the 8. This number is nearer to 3.48 than it is to 3.47, so the correct answer, to 2 d.p. is **3.48**

Example 2

The result of a calculation comes to 7.42 Give this to one d.p.

This number is nearer to 7.4 than it is to 7.5, so the correct answer, to 1 d.p. is **7.4**

(Help) desk

If the next decimal place along is five or more, round up.
If the next decimal place along is less than five, round down.

Now try calculating with decimals.

Question 1

Alan buys a burger for £1.75, a cold drink for £0.80, and a chocolate bar for £0.85

What is the total cost?

> Money is always written to two decimal places with a pound sign.

    Answer: £3.40

Question 2

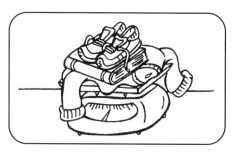

Lisa weighed several items to make sure she would not exceed her baggage allowance for flying.

shoes	600 grams
jacket	430 grams
jumper	510 grams
book	350 grams

(i)   What was the total weight of these items in grams?

(ii)   What was the weight of the items in kilograms?

      (1 kg = 1000 g)

(iii) Round off your answer to (ii) to one decimal place

(iv) Round your answer to (iii) to the nearest kilogram.

Answers:

(i)   1 890 grams

(ii)   1.89 kg

(iii)   1.9 kg

(iv)   2 kg

Help desk

To change grams to kilograms:

divide by 1000

Rounding to the **nearest number**:

round up if the next digit is 5 or more

round down if the next digit is less than 5

Help desk

Fractions divide things into parts.

The top number is how many parts we have.

The bottom number is how many parts there are altogether.

The diagram shows $\frac{1}{6}$

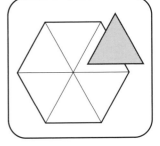

Fractions

We are used to the decimal system for our weights and measures and for our money. For example ten pence can be written as £0.10

There are 100 pence in one pound, so ten pence is ten hundredths of a pound.

We can write the number as a fraction, $\frac{10}{100}$

Fractions should always be written in their *lowest form*, using the smallest numbers.

To find the lowest form, think of a number that divides into both the top and bottom of the fraction.

The fraction is $\frac{10}{100}$ so divide both the top and bottom numbers by 10.

$$\frac{\cancel{10}}{\cancel{100}} = \frac{1}{10}$$

Here is another example. $\frac{75}{100}$

Try dividing both the top and the bottom numbers by 5.

$$\frac{\cancel{75}}{\cancel{100}} = \frac{15}{20}$$

This is still not the lowest form. Try 5 again.

$$\frac{\cancel{15}}{\cancel{20}} = \frac{3}{4}$$

The fraction is now in its lowest form.

Questions:

Give these fractions in their lowest form. For each one, find a number that divides into both the top and the bottom. Then cancel down until you get to the smallest numbers.

$$\frac{4}{8} \quad \frac{3}{12} \quad \frac{14}{16} \quad \frac{2}{10} \quad \frac{24}{36}$$

Answers:

$\frac{1}{2}$ (Divide by 2) $\frac{1}{4}$ (Divide by 3)

$\frac{7}{8}$ (Divide by 2) $\frac{1}{5}$ (Divide by 2)

$\frac{2}{3}$ (Divide by 12, or cancel down in stages dividing by 2 then 2 then 3).

To change a fraction to a decimal number

- Decide how many decimal places you need in your answer
- Put a decimal point after the top number in the fraction
- Add zeros for as many decimal places as you need plus one more
- Divide the bottom number into the top number
- Round up or down to the correct number of decimal places.

Example

$$\frac{4}{8} = 4 \div 8 = 0.50$$

Now try these fractions for yourself. Give all the answers to 3 decimal places.

$$\frac{3}{4} \quad \frac{1}{6} \quad \frac{5}{8} \quad \frac{2}{5}$$

Answers:

0.750   0.167 (rounded to 3 d.p.)   0.625   0.400

Help desk

Always add zeros to make the correct number of decimal places.
Always put in the zeros, if a number 'won't go' when you are dividing.

Percentages

Percent means "out of a hundred", so if we have ten out of a hundred, we have 10%

Now we have several different ways to write the same number.

0.10	$\dfrac{10}{100}$	10%
decimal number	fraction	percentage

To change a decimal number to a percentage, just multiply by 100.
(Quick tip – move the decimal point 2 places to the right)

To change a percentage to a decimal number, just divide by 100.
(Quick tip – move the decimal point 2 places to the left)

Questions

(i) Complete the table below with the percentages.

decimals	0.10	0.20	0.25	0.50	0.75	1.00
%						

(ii) Now try to give each of these as a decimal number.

   5%   15%   30%   40%   85%   100%

Answers:
(i) 10%   20%   25%   50%   75%   100%
(ii) 0.05   0.15   0.30   0.40   0.85   1.00

Now look at this example.

Kim pays 5% VAT on her gas bill. The charge for gas is £35. How much does Kim pay in VAT?

Work out $\dfrac{5}{100}$ of £35

$$\frac{5}{100} \times 35 = 175 \div 100 = 1.75$$

Answer: Kim pays £1.75 in VAT.

Try these questions for yourself using the methods shown above.

(i)   Eric books a holiday. The price is £850. He pays the travel agent a deposit of 20%.

How much deposit does Eric pay?

(ii) Sam buys a car, price £6500. Her first payment is 40% of the full price.

She pays the remainder in equal amounts each month for 5 months.

How much money is 40% of the full price?

How much does Sam pay each month?

(iii) A company has total annual costs of £200 000. They are divided as follows.

Wages	60%
Buildings	5%
Materials	15%
Running costs	20%

How much is each type of cost in money?

What fraction of the total annual costs is wages?

(iv) Three quarters of the tickets for a football match were sold to supporters of the home team.

What percentage of the crowd supported the home side?

Answers:

(i) £170

$$\left(\frac{20}{100} \times 100\right)$$

(ii) £2 600

$$\left(\frac{40}{100} \times 6500\right)$$

£780 each month   6500 – 2600 = 3 900

3 900 ÷ 5 = 780

(iii) £120 000
  £10 000
  £3 000
  £40 000
  Wages are $\frac{3}{5}$ (Divide the top and bottom by 10, then by 2)

(iv) 75%

Ratios

Ratios are used to compare quantities. The ratio shows the proportions of each quantity.

Example 1

A room is 4 metres long by 3 metres wide. What is the ratio of its length to its width?

> **Answer:** The ratio of length to width is four to three. In numbers this is 4 : 3

Help desk

The ratio is given in simple numbers.

The numbers show the size of one quantity compared with the size of the other.

The numbers must always be written in the same order as the names of the quantities.

Example 2

There are 20 children in class A and 30 children in class B. What is the ratio of A to B?

> **Answer:** The ratio of A to B is 20 : 30 but it is better to give this as 2 : 3

The numbers can be simplified by dividing both numbers as we did with fractions.

20 : 30 is the same as 2 : 3 (divide both numbers by 10).

Try these questions.

(i) Two large boxes contain fruit. There are 50 apples in one box and 40 pears in the other.
What is the ratio of apples to pears?

(ii) A farmer has five times as many sheep as cows. What is the ratio of cows to sheep?

(iii) A travel company keeps records of where its customers go for holidays. In May, 600 people went to Spain and 400 people went to Greece. What was the ratio of Spanish holidays to Greek holidays?

Answers:

(i) 5 : 4

> The ratio is 50 : 40. Divide both numbers by 10

(ii) Cows to sheep = 1 : 5

> The numbers must be in the same order as the names of the animals. For every 1 cow there are 5 sheep.

(iii) 3 : 2

> The ratio is 600 : 400.
> Divide by 100 = 6 : 4.
> Divide by 2 = 3 : 2

Measuring with scales

There are three things to check when you use a scale:

- What are the units of measurement?
- What is the value of each major division?
- What is the value of each minor division?

Help desk

major division – the space between the big marks that are numbered

minor division – the space between the little marks that are not numbered

Example 1

The scale on the top of the next page measures in **inches**. Each major division represents **one inch**. Each minor division is **one tenth**, (0.1) of an inch. (Count the *spaces* between the marks).

The pointer shows a length of **2.3 inches**.

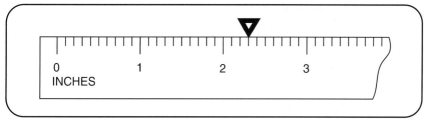

Example 2

The scale below measures in **centimetres**. Each major division represents **10 cm**. Only five minor divisions are shown between the major divisions. Each minor division is therefore worth **2 cm**.

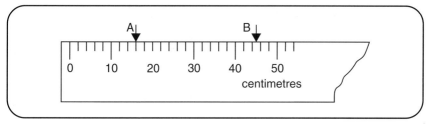

Arrow A shows a length of **16 cm**. (We have to count two for each minor division; so 10, 12, 14, 16).

Arrow B shows a length of **45 cm**. (Half way between the marks counts one; so 40, 42, 45).

Try reading the scales below:

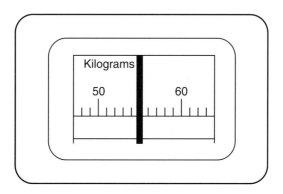

(i) How much does the person on these bathroom scales weigh?

31

(ii) What is the temperature on this medical thermometer?

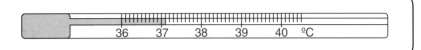

(iii) How tall is this person, to the nearest centimetre?

Answers:
(i) 55 Kilograms
(ii) 37.1°C
(iii) 148 cm

Areas and volumes

You need to remember the formulae for finding:
- The area of a rectangle
- The volume of a rectangular box or room.

Help desk

Area – the space inside a flat shape
Volume – the space inside a container

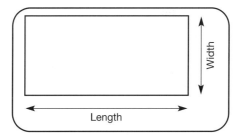

Area = length × width $A = l \times w$

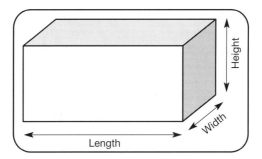

Volume = length × width × height $V = l \times w \times h$

Remember:
multiply (×) the measurements, don't add them by mistake.

When you calculate area and volume, you must always give the units of measurement.

Area is always square units, shown by a small ^{2} e.g. square metres is written as m^2
Volume is always cubic units, shown by a small ^{3} e.g. cubic metres is written as m^3.

Example:

A room is 7 metres long, 3.5 metres wide and 3 metres high.

What is the area of the floor?

What is the volume of the room?

   Answers:

   $A = l \times w$

   $A = 7 \times 3.5 = \textbf{24.5 m}^2$ (24.5 square metres)

$$V = l \times w \times h$$
$$V = 7 \times 3.5 \times 3 = \mathbf{73.5\,m^3} \quad \text{(73.5 cubic metres)}$$

Try these questions.

(i) The back of Bob's van measures 4 metres long by 2 metres wide by 1.5 metres high.
What is the area of the floor?
What volume of space is there in the back of the van?

(ii) Megan wants to put fertiliser on her lawn. The instructions say "use 3.5 grams per square metre."
The lawn is 8 metres long by 5 metres wide.
How much fertiliser does Megan use?

(iii) One crate of drinks measures $50\,cm \times 50\,cm \times 50\,cm$.
The crates are piled in a stack. The stack is 4 crates long, 2 crates wide and 5 crates high.
What volume of space does the whole stack occupy?
(Note: If you multiply in centimetres you will get a very large number, $5{,}000{,}000\,cm^3$
It is better and safer to work out the measurements of the stack, convert them from centimetres to metres; then do the multiplying).

Answers:

(i) Area of floor $= 8\,m^2$ Volume of space $= 12\,m^3$

(ii) Amount of fertiliser $= 140$ grams.
(Area of lawn $= 8 \times 5 = 40\,m^2$. Then multiply
$40 \times 3.5 = 140$)

(iii) Volume of stack $= 5\,m^3$
length $= 50 \times 4 = 200\,cm = 2$ metres
width $= 50 \times 2 = 100\,cm = 1$ metre
height $= 50 \times 5 = 250\,cm = 2.5$ metres
$\therefore$ Volume of stack $= 2 \times 1 \times 2.5 = 5\,m^2$

$\boxed{\textbf{Help}}\,d\,e\,s\,k$

There are 100 centimetres in 1 metre.
If this measurement is cubed to find a volume, we multiply
$100 \times 100 \times 100$
So there are 1 million cubic centimetres in 1 cubic metre

Working with statistics

Mean and Range

The **mean** number is a kind of average.

By calculating the mean we can use one number to represent a whole group, even if all the numbers in the group are different.

To find the mean:

- Add all the numbers together
- Divide by the number of numbers

Example:

A golfer made these scores in a competition:

68   70   71   69

What was the golfer's mean score?

> **Answer:**
>
> 68 + 70 + 71 + 69 = 278
>
> 278 ÷ 4 = 69.5
>
> The golfer's mean score is **69.5**
>
> 69.5 is not in the list. It is somewhere between the highest and lowest numbers. It represents the group of numbers. The mean is a useful number when we want to compare groups of numbers.

The **range** is the spread of numbers.

To find the range, we calculate the difference between the highest and lowest numbers.

- Subtract the lowest number from the highest number.

Example:

Here are the heights of ten children, measured in centimetres.

126	119	127	115	116
117	120	127	121	125

What was the range of their heights?

> **Answer:**
>
> The highest is 127cm. The lowest is 115cm.
>
> 127 – 115 = 12
>
> The range is **12 cm**.

Try these questions:
(i) Eggs are graded by weight. The weights of ten eggs, in grams, are shown below.

53 50 49 47 56
56 51 55 49 45

Calculate the mean weight, in grams, of the eggs.
Calculate the range of the weights.

(ii) In five matches a cricket team made these scores:

First innings	175	204	232	196	177
Second innings	213	211	238	172	212

What was the mean score for all 10 innings?
What was the range of all the scores?

(iii) Two athletes were training for the 100 m sprint. They each recorded their times in five training sessions.

First athlete	10.5 sec	10.6 sec	10.5 sec	9.9 sec	10.2 sec
Second athlete	10.2 sec	10.1 sec	10.0 sec	9.9 sec	10.0 sec

What was the mean time for each athlete?
What was the range of times for each athlete?

Answers:
(i) Mean = 51.1 g (511 ∏ 10)
    Range = 9 g (56 – 45)
(ii) Mean score = 203 (2030 ∏ 10)
    Range = 66 (238 – 172)
(iii) First athlete:
        Mean time = 10.34 seconds (51.7 ∏ 5).
        Range = 0.7 seconds (10.6 – 9.9)
    Second athlete:
        Mean time = 10.04 seconds (50.2 ∏ 5)
        Range = 0.3 seconds (10.2 – 9.9)

Help desk

Did you remember to give the units of measurement with your answers?

Read and interpret information

Tables

Tables are a good way to show numerical information. The numbers are lined up in rows and columns. This makes it easy to find a particular number by referring to the row and column headings.

Some tables of different types are shown in the questions below.

Question 1:

Choosing which Building Society account to open.

Account name	Minimum investment	Interest rate %
Instant Access	£1 £100	0.50 4.75
Young Saver	£1 £10	1.00 4.75
Savings Bond	£1 £500	1.00 5.00

(i)   You want an Instant Access account, and you have £100 to invest. What interest rate will you get?

(ii) You want the highest possible interest rate, and you have £500 to invest. Which account should you open?

(iii) You are a "Young Saver". You want to open an account with £5. What interest rate will you get?

Answers:

(i)   4.75%

(ii) Savings Bond.

(iii) 1.00%.

The numbers you were given in the question are not in the table. You have to **interpret** the table. Make use of the numbers that are there, to find out the information you want. £5 is more than £1 but the next 'minimum' is £10, so you must use the lower interest rate.

Question 2:

Here is part of a train timetable.

Times are given using the 24-hour clock. The numbers show the hours then the minutes.

Poole	0541	0641	0710	0750	0810	0850
Parkstone	0545	0645	0714	0754	0814	0854
Branksome	0549	0649	0717	0758	0817	0858
Bournemouth	0554	0654	0722	0803	0822	0903

(i)   You want to get to Bournemouth by 7 o'clock in the morning. What time does a suitable train leave Parkstone?

(ii) If you catch the 0810 from Poole to Bournemouth, how long does the train journey take?

(iii) Does the journey take the same time if you catch the 0850 from Poole to Bournemouth?

Answers:

(i)   0645.

   Find the rows for Bournemouth and Parkstone, and the column with the Bournemouth time before 0700 (7 o'clock in the morning). 0654 is just before 0700, so the second train would be suitable.

(ii) 12 minutes

   For this question you need to interpret the table by doing a calculation. The train leaves at 0810. That is eight hours and ten minutes, or 10 minutes past eight. It arrives at 22 minutes past eight. So the journey time is 22 – 10 = 12.

(iii) No. (It takes 13 minutes).

   Count the minutes from 0850 to 9 o'clock, then add the minutes after 9 o'clock.

Charts

There are many different types of chart. We shall look at a bar chart, a pie chart and a tally chart as examples.

   Here is a **BAR CHART** showing the results of a survey to find out how shoppers use the town car park.

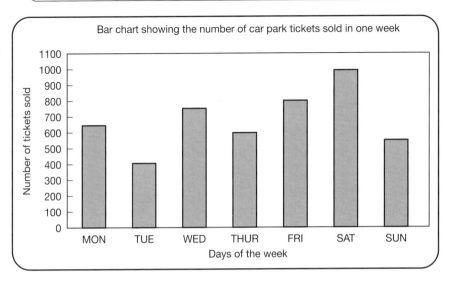

Bar chart showing the number of car park tickets sold in one week

First look at the information given in the chart. To make sense of a chart, you need **all** this information:

- A title, telling you what the chart is about
- A horizontal axis along the bottom, with names for each bar and a label telling you what the bars represent
- A vertical scale up the left hand side, with the numbers equally spaced and a label telling you what they represent.

Now try to answer these questions.

(i) On which day were most tickets sold?

(ii) How many tickets were sold in total, during the whole week?

(iii) How many people bought tickets on Friday, Saturday and Sunday, in total?

(iv) What is the average (mean) number of tickets sold in a day during the week shown?

(v) What is the range of tickets sold during the week shown?

Answers:

(i) Saturday

> just look for the highest bar

(ii) 4 750.

> If you got a different answer, check the heights of the bars again. You need to read between the lines for Monday, Wednesday and Sunday. Use a ruler to draw a line back to the scale, if this helps.

(iii) 2350.

First make sure you select the right bars, then do the addition.

(iv) 678

To find the mean, you add all the numbers together and then divide by the number of days in the week 4750 ÷ 7 = 678.57 (Remember you should have done this in your head, and only checked it with a calculator).

We have rounded the number off to a whole number. Can you say why? You cannot have 0.57 of a ticket, so we have to round **down** to the next **whole** ticket.

(v) 600

The range is the **difference** between the highest and the lowest numbers. You have to do the calculation, 1000 – 400 = 600 and give the answer. Do not just give the range as 1000 – 400 or as 400-1000

Pie charts

Pie charts are used to show proportions in fractions or percentages.

Look at the pie chart below. It was published by a local council to show taxpayers where the council money comes from.

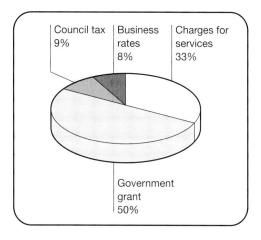

Council tax 9%

Business rates 8%

Charges for services 33%

Government grant 50%

Try these questions:
(i)   Make a table showing all the information given in the pie chart.
(ii) What fraction of the council's income is Government grant?
(iii) What fraction, approximately, of the council's income is Charges for services?
(iv) Approximately what fraction of the whole amount do Council tax and Business rates add up to?

Answers:
(i)   Your table should look something like this

Source of income	Percentage %
Charges for services	33
Government grant	50
Council Tax	9
Business Rates	8

(ii) half, $\frac{1}{2}$

> You can judge this by looking at the chart

(iii) approximately one third, $\frac{1}{3}$

> This might be harder to judge, but you can check by calculation.
> Remember, "percent" means the number of parts we have out of 100.
> 100% is the whole amount.
> $\frac{33}{100}$ will nearly cancel down to $\frac{1}{3}$
> (33 divides into 100 just over 3 times).

(iv) one sixth, $\frac{1}{6}$

> Work it out this way: $9 + 8 = 17$
> 17 goes into 100 nearly six times, so the answer is one sixth

Tally charts

Tally charts are a good way to record a count. Look at the example below.
The chart shows the number of people who visited a tourist attraction in one morning. As each visitor bought a ticket, a

stroke of one was recorded. The strokes were recorded in groups of five.

~~HH~~ represents five people

Time (am)	Tally of visitors	Total number (the frequency)
9.00 – 9.30	IIII	
9.30 – 10.00	HH II	
10.00 – 10.30	HH HH HH HH IIII	
10.30 – 11.00	HH HH HH	
11.00 – 11.30	HH HH HH HH	
11.30 – 12.00	HH HH HH HH HH HH	

Now answer these questions by looking at the chart.
(i)   What are the totals for each time period (the frequency)?
(ii) How many people bought tickets between 10.00 and 10.30?
(iii) How many **more** people bought tickets in the 11.30 – 12.00 time period, than between 10.00 and 10.30?
(iv) What was the total number of visitors in the whole morning?
(v)   What fraction of the total number of visitors bought tickets after 11 o'clock?
(vi) What percentage of people bought tickets between 10.30 and 11.00?

Answers
(i)   4,   7,   24,   15,   20,   30
(ii) 24
(iii) 6   (30 – 24 = 6)
(iv) 100
(v) $\frac{1}{2}$

$\frac{50}{100}$ cancels down to $\frac{1}{2}$ by dividing top and bottom by 50

(vi) 15%

$\frac{15}{100} \times 100$ %

Diagrams

The diagrams you are most likely to meet in tests are plans or maps. To interpret them you must be able to use a **scale**. Scales can be given in words or written as a ratio. For example,

> scale: 1 cm to 1 m means that one centimetre on the drawing represents one metre in the actual size.
> scale 1: 50 means that one unit on the drawing represents 50 of the *same units* in the actual size.

You may need to convert units to get a sensible answer. Look at this map, then try the questions below.

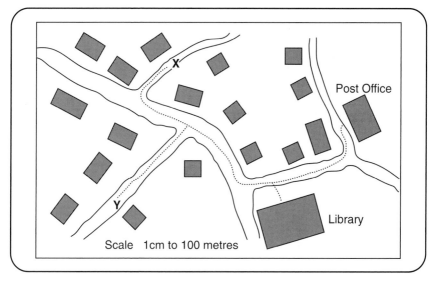

Scale   1cm to 100 metres

Help desk

If you use a ruler to measure on a map like this, it is difficult to be accurate.

Measure in centimetres.

Round off the number so that you have an approximate measurement.

Multiply by the scale factor.

Give an approximate answer.

Try these questions.

(i) How far is it from house X to the Library?

(ii) How far is it from house Y to the Post Office?

(iii) Write the scale as a ratio.

Answers:

(i)   750 m   (to the nearest 50 metres)

(ii) 950 m   (to the nearest 50 metres)

(iii) 1 : 10 000

Remember, when the scale is given as a ratio, both numbers must be in the same units.

100 metres = 100 × 100 cm. Therefore 1 cm on the map represents 10 000 cm on the ground

---

Line graphs

Help desk

Graph – points plotted from two axes.

The axes have scales on them.

A line or curve shows the pattern and direction in which things are changing.

Graphs are easy to read if you make use of **all** the information you are given.

Look at the temperature conversion graph at the top of the next page.

The title tells us what the graph shows.

The horizontal scale is labelled "Degrees Celsius" (°C).

The scale has intervals marked at every 50 degrees, from 0 °C to 300 °C.

There are gridlines exactly half way between the marked values. These must be worth 25 degrees each. So we can work out the complete scale by counting in 25s.

The vertical scale is labelled "Degrees Fahrenheit" (°F).

This scale has intervals marked at every 100 degrees from 0 °F to 700 °F.

The gridlines in between are each worth 20 degrees.

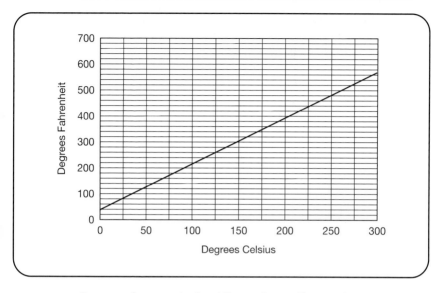

Once you have worked out the scales on the graph you can read off any pair of values.

Example:
If the Celsius temperature is 150 °C, what is the Fahrenheit temperature?

Answer:
About 300 °F.

Find 150 on the Celsius scale. Read up to the graph line and across to the Fahrenheit scale. Read the number in °F. Don't forget to give the units.

Now answer these questions.
(i) A pie has to be cooked at 400 °F. What is the nearest Celsius temperature?
(ii) The weather forecast predicts a temperature of 25 °C. How warm is this in °F?
(iii) Water boils at 100 °C. Estimate how hot this is in °F.

Answers:
(i) 200 °C
(ii) 80 °F
(iii) Estimate = 210 °F

You might know the accurate figure is 212°.

4 Practice Your skills at Level 1

Introduction

The test has 40 multiple-choice questions.

In each question there is a context.

The text of the question describes a situation, or gives you information about a person or organisation.

You are then given four possible answers: A, B, C and D.

You have to choose the correct option, then mark the letter of your choice on an answer grid.

DON'T JUST GUESS – WORK IT OUT

Don't expect to be able to work it out in your head without writing anything down.

You should be given some rough paper on which to do the calculations. If not, ask for some. If there isn't any, then use the margins in the question booklet.

How to tackle the paper

Work through the paper, attempting all the questions, in order. If you do not understand one of the questions, mark it on the paper, (e.g. with a circle or star), and go on to the next question. When you get to the end of the paper, go back to the beginning. Read through again, looking for any you missed on the first run through.

Have another go.

If you can't decide which calculation to do,

Try to think of the scene being described – what sort of number are you looking for?

- If you expect the answer to be a large number, you need to add or multiply some of the numbers you are given
- If you expect the answer to be a small number, you need to subtract or divide some of the numbers you are given

Try to judge the size of the answer – should it be units, tens, hundreds, etc.

If all else fails it is better to guess than to leave a blank. You have a 25% chance of guessing correctly. That's better than the lottery!

 If you still have time, check through all your answers again.

How to tackle each question

Read through all of the information, so that you know what the question is about.

Read the sentence that contains the actual question carefully. Pick out the key words. Underline them if it helps to focus your mind.

Go back to the information and find the numbers you need for your calculation.

Write out the calculation.

Check your result with the options you are given: A, B, C and D, and decide which letter gives the correct answer.

Mark this letter on the answer grid.

If you can't find your result among the options, A, B, C or D

You made a mistake! Check:

- Did you do the correct calculation?
  (e.g. if you added two numbers, should you have multiplied them?)
- Did you carry out the calculation correctly?
  (e.g. did you remember your multiplication tables correctly?)

Go through the question again, or, if you think you are short of time, go on with the rest of the test, but mark this question to check again at the end.

Remember, the *wrong* options are the numbers you get if you make a mistake in your calculations.
Nasty!

Use all the time you have.
Check each question.
Check the whole paper.

In the next section you will find a sample test paper. There are 20 multiple choice questions. These are all based on questions taken from past papers at Level 1. You should be able to complete all 20 in half an hour.

Do the paper in test conditions.

Sit somewhere quiet.
Give yourself half an hour.
Make sure you have some rough paper to do the calculations on.
No calculator!
No asking for help!

Good luck.

SAMPLE TEST PAPER

Questions 1 and 2 are both about football clubs.

The table gives ground sizes and crowd capacities for six football clubs in 1998.

Football club	Ground size Width (m)	Length (m)	Crowd capacity
Arsenal	61	92	38 500
Aston Villa	60	96	39 217
Bradford City	61	92	18 018
Leeds	65	105	40 205
Liverpool	62	92	45 362
Manchester United	65	98	56 387

1   Which football club's ground had the largest **area**?
   A   Manchester United
   B   Liverpool
   C   Leeds
   D   Bradford City

2   Which number is the **combined** crowd capacity at Liverpool **and** Manchester United?
   A   One hundred thousand, six hundred and forty-nine
   B   One hundred and one thousand, six hundred and forty-nine
   C   One hundred and one thousand, seven hundred and forty-nine
   D   One million and one thousand, seven hundred and forty-nine

3   Ajay works out 6 000 ÷ 19 on his calculator.
This is the display on the calculator.

$$315.789473$$

Which option shows this answer to the **nearest whole number**?

A   316
B   315
C   310
D   300

4   Breakfast cereal is usually served with milk.
Which unit is used for measuring milk?
A   cm²
B   ml
C   mm
D   km

5   A portion of breakfast cereal contains 25% of the Recommended Daily Allowance (RDA) of Vitamin B.

Which option shows the number of portions of the cereal needed for the full RDA of Vitamin B?
A   4
B   5
C   20
D   25

6   Plain chocolate contains 1% water, 5% protein, 29% fat and the rest is carbohydrate.

Which is the percentage of carbohydrate in plain chocolate?
A   35%
B   45%
C   55%
D   65%

7   A 100 gram portion of kipper fillet contains 18 grams of protein.

Which option shows the amount of protein in a 250 gram portion of kipper fillet?
A   45 g
B   36 g
C   25 g
D   7.2 g

8   According to the rules, a table tennis ball must weigh
    2.5 units.

    Which is the correct unit?

    A   centimetre
    B   gram
    C   litre
    D   square metre

9   This graph shows speeds in metres per second, and
    equivalent speeds in miles per hour.

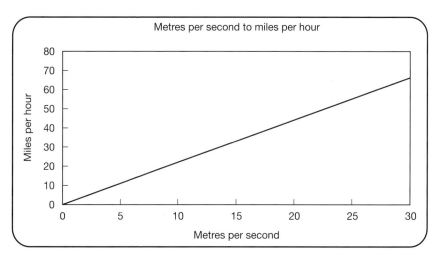

Metres per second to miles per hour

    Bungee Jumpers can fall at speeds as fast as 25 metres
    per second.

    Use the graph to find this speed in miles per hour.

    Which option gives this speed to the **nearest five miles
    per hour**?

    A   10 miles per hour
    B   45 miles per hour
    C   55 miles per hour
    D   60 miles per hour

10 The ratio of a piece of bungee cord stretched to its fully
    stretched length is 1 : 3

    The length of a slack bungee cord is 9 metres.

Which option shows its fully stretched length?

A   36 m
B   27 m
C   12 m
D   9 m

11 The scale reading shows the weight
of a bag of coins.

Which option shows the correct weight?
A   500 g
B   520 g
C   550 g
D   580 g

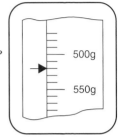

12 In a restaurant, lunch prices are £5 for an adult's meal and
£2.50 for a child's meal.

Which amount is the total cost of lunch for a group of 20
adults and 10 children?

A   £75
B   £125
C   £150
D   £225

Questions 13 and 14 are about a water company's costs.

The water company sent its customers this chart to show how
it spent its money.

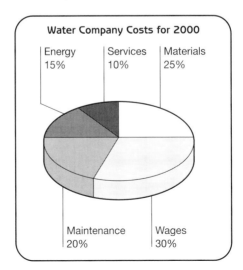

Water Company Costs for 2000

| Energy 15% | Services 10% | Materials 25% |

| Maintenance 20% | Wages 30% |

13 From the information given by the chart, which of the
following statements is true?

 A   maintenance costs more than services
 B   energy costs more than materials
 C   services cost more than energy
 D   materials cost more than wages

14 Which fraction of the company's costs were materials?

 A $\frac{1}{25}$

 B $\frac{1}{10}$

 C $\frac{1}{5}$

 D $\frac{1}{4}$

15 In one week a salesperson drives 540 kilometres.

 The car uses 34 litres of petrol.

 One litre of petrol costs £0.85

 Which calculation shows the car's performance in
 kilometres per litre ?

 A   540 ÷ 0.85
 B   540 ÷ 34
 C   540 × 34
 D   540 × 0.85

Questions 16 and 17 are about the water tank shown in the
diagram.

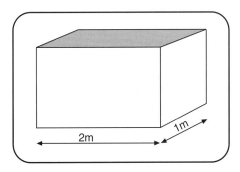

When the tank is full, the water is 1 metre deep.

16 Which is the **volume** of water in the tank when it is full?

A $1\,m^3$

B $2\,m^3$

C $3\,m^3$

D $4\,m^3$

17 One quarter of the water from a full tank is used.
Which option shows the depth of the water **left in the tank**?

A $0.10\,m$

B $0.25\,m$

C $0.50\,m$

D $0.75\,m$

18 A petrol station has this sign beside the pumps.

MOBILE PHONE TOP-UP VOUCHERS
1 point for every £6 spent on petrol
Collect 25 points for one voucher

A customer spends £30 each week on petrol, and collects the points.

Which option shows how long it takes this customer to collect enough points for one voucher?

A   3 weeks

B   4 weeks

C   5 weeks

D   6 weeks

Questions 19 and 20 are about an estate agent.

19 An estate agent values a house at one million, two hundred and fifty thousand pounds.

Which figure shows this price?

A   £100 250

B   £125 000

C   £1 000 250

D   £1 250 000

20 The estate agent gets 2% commission on each sale.

Which is the commission on a house that sells for
£94 000?

A   £470

B   £940

C   £1 800

D   £2 000

END OF TEST

5

Apply Your Skills at Level 1

Understanding Part B

Part B of the specifications is about **using** numbers in a job or activity at Level 1.

You will be given some activities that involve working with numbers. You must keep the work that you do for these activities in a "portfolio".

Your **portfolio** is the collection of work that proves you can **apply** skills and knowledge to do with numbers.

Part B is divided into three sections. These are the three stages in solving problems and giving people information that depends on numbers.

N1.1 is about **obtaining** and **interpreting information**
N1.2 is about **calculating**
N1.3 is about **interpreting results** and **presenting findings**

On the next few pages you will find an example of an activity that could contribute to a Level 1 Portfolio. The activity has been worked through for you.

What you should do next

Read through the Activity Brief. This describes the **purpose** of the activity.

Work through the **Activity Plan** for yourself.

Use the **Evidence Check** to compare your work with the example given.

Read through **Check the Spec** to see what has been achieved through this activity.

Note what still has to be done to achieve the whole unit.

Help desk

Purpose – Points to keep in mind as you do an activity.

What are you being asked to find out?

Who needs the information?

What do they want it for?

How can you make sure that you provide accurate information?

In this activity, the **purpose** is to help the Smiths choose the best way to go on holiday.

ACTIVITY BRIEF

The Smith family from Bournemouth are planning a holiday in Ireland. They want to visit Dublin, Cork and Killarney. They have up to 7 days available.

Their options are:

1. **Fly to Dublin from Bournemouth, and hire a car**

2. **Drive their own car all the way, using the ferry from Fishguard to Rosslare.**

There are two adults and two children in the family.

They will arrange their own accommodation.

They do not like driving at night.

The Smiths want to take the cheapest option, but they also want to spend as much time as possible in Ireland.

You are asked to obtain the relevant information, do the appropriate calculations and make a recommendation for which option they should choose.

Some information sheets are supplied. These contain:

- Distance charts for car travel in Great Britain and in Ireland
- A map of Great Britain and Ireland
- Price tables for air travel between Bournemouth and Dublin, and ferry crossings between Fishguard and Rosslare
- Car hire costs and petrol costs

Details of what you need to do are given in the **Activity Plan**.

Help desk

Activity Plan.

You must have a plan of how you are going to do the activity.

The plan is a detailed set of step-by-step instructions that take you through the whole activity.

At Level 1, your teacher or assessor can give you the Activity Plan.

ROAD DISTANCES IN IRELAND

ARMAGH	99 53 54 40 61 236 78 81 168 50 148 152 240 63 171 71 87 19 36 129 99 92 187 236 178 164
ATHLONE	152 153 139 153 135 112 77 68 79 57 75 142 162 74 140 31 101 106 44 20 72 87 138 104 114
BALLYMENA	42 29 28 283 90 124 221 82 202 195 288 21 219 52 129 58 55 175 153 124 234 284 221 207
BANGOR	14 71 276 124 117 223 97 202 189 281 37 213 87 122 52 82 169 154 140 227 277 214 201
BELFAST	58 263 110 104 210 84 189 175 267 23 199 73 109 38 69 155 140 126 214 263 201 187
COLERAINE	290 77 153 222 91 205 225 294 49 226 32 141 88 64 183 154 117 241 290 250 237
CORK	249 159 86 213 128 92 55 285 63 274 150 225 240 108 156 209 53 75 78 116
DONEGAL	145 163 36 128 187 254 115 186 44 122 96 41 152 92 40 210 250 216 226
DUBLIN	142 108 136 71 188 126 119 145 50 66 111 51 90 132 110 184 97 82
ENNIS	147 43 94 91 232 22 207 101 172 174 91 87 124 47 87 100 138
ENNISKILLEN	119 150 219 106 151 61 78 69 27 119 70 42 164 215 180 191
GALWAY	105 133 210 65 172 90 161 146 93 48 88 89 129 135 155
KILKENNY	122 197 71 208 73 137 174 31 95 147 47 134 30 50
KILLARNEY	290 68 281 159 229 246 137 162 214 85 20 119 158
LARNE	223 76 133 62 75 179 164 150 238 287 225 211
LIMERICK	212 90 161 178 68 94 146 24 64 77 116
LONDONDERRY	136 89 34 178 131 83 225 276 238 227
MULLINGAR	71 102 42 40 82 101 154 102 112
NEWRY	55 117 111 111 176 225 163 149
OMAGH	144 97 69 191 242 204 193
PORT LAOISE	64 116 59 132 60 70
ROSCOMMON	52 107 158 124 134
SLIGO	170 210 176 186
TIPPERARY	81 53 91
TRALEE	131 169
WATERFORD	38
WEXFORD	

ROAD DISTANCES IN GREAT BRITAIN

LONDON	548
ABERDEEN	238 471
ABERYSTWYTH	419 191 341
AYR	216 607 222 465
BARNSTAPLE	352 188 324 131 455
BERWICK	120 433 123 291 178 274
BIRMINGHAM	244 337 167 193 303 189 129
BLACKPOOL	107 579 213 437 121 417 161 275
BOURNEMOUTH	60 609 290 468 203 416 171 305 95
BRIGHTON	120 517 132 375 100 365 88 213 82 170
BRISTOL	60 465 231 352 267 296 113 229 159 121 171
CAMBRIDGE	155 537 118 395 137 385 108 233 128 205 47 206
CARDIFF	313 234 236 92 372 87 198 102 344 374 282 256 302
CARLISLE	79 587 322 491 275 418 205 329 182 81 209 121 244 398
DOVER	413 126 335 82 471 58 298 201 444 473 381 336 401 98 458
EDINBURGH	200 591 206 449 55 439 162 287 84 175 84 251 121 356 248 455
EXETER	260 527 56 385 243 380 177 223 233 311 153 312 111 292 350 392 227
FISHGUARD	524 158 446 136 582 192 409 312 555 584 492 469 513 210 608 133 566 502
FORT WILLIAM	413 149 336 35 472 105 298 201 444 474 382 358 402 99 498 46 456 392 102
GLASGOW	103 482 111 341 126 330 54 178 102 156 36 123 65 247 195 347 109 170 458 347
GLOUCESTER	82 551 309 437 313 382 192 316 206 133 216 69 252 344 133 422 296 357 555 444 192
HARWICH	282 463 105 321 341 345 167 159 312 342 251 275 205 227 367 327 324 161 438 327 216 354
HOLLYHEAD	573 107 496 216 632 219 458 362 604 634 542 519 562 259 658 157 616 552 65 174 507 604 468
INVERNESS	268 286 190 145 326 139 153 56 299 328 237 253 257 51 352 151 310 246 262 151 202 340 182 311
KENDAL	188 361 229 264 321 187 140 142 284 283 232 140 252 171 261 232 305 285 382 271 197 226 220 431 165
KINGSTON UPON HULL	198 328 174 215 310 159 121 87 264 263 220 148 240 122 270 199 294 230 332 222 185 234 166 382 71 61
LEEDS	

HOLIDAYS IN IRELAND – PRICE INFORMATION

Flights between Bournemouth and Dublin

Basic price for a return ticket: £150
Reductions of up to 40% are available if you book early!

Flight times:

Dep. Bournemouth 13.45	Arr. Dublin 14.55
Dep. Dublin 12.10	Arr. Bournemouth 13.20

Check in one hour before departure time.

Ferry between Fishguard and Rosslare (Prices are in £ sterling)

	one way	midweek return	5 day return
car + driver	144	249	184
car, driver + 1 passenger	154	269	204
special family rate	159	289	219
car, driver + 4 passengers	164	294	224

Sailing times
Dep. from both ports at 08.00 and 13.30.
Crossing time 2 hours.
Check in one hour before departure time.

■ NOTE: To estimate how long it takes to reach the ferry port, travellers should allow for an average speed of 40 mph.

Car Hire in Ireland – Rates (Prices are in £ stirling)

	daily price	weekly price
small family car	18.00	105.00
medium family car	21.00	125.00
8 seater	32.00	195.00

Unlimited mileage – customers pay for all fuel used.

■ NOTE: Drivers should allow 10p per mile for fuel when estimating cost of travel.

ACTIVITY PLAN

Option 1

Look in the price tables to find the cost of a return flight air ticket between Bournemouth and Dublin.

Calculate the price with a 40% reduction for booking early.

Work out what the Smith family of four would have to pay if they book early and get the reduction. Record the answer in pounds and pence.

Check the flight times. How long will the Smiths have in Ireland? For how long will they need a car?

Look in the price tables to find the cost of car hire.
Calculate the cost for one week. Record it in pounds and pence.

Look at the map of Ireland to work out a route between the towns that the Smiths want to visit, starting and ending in Dublin.

Look in the distance chart for Ireland to find the distances between the towns.

Work out the total miles in Ireland.

Look up the petrol price, and work out the cost for the total distance. Record the answer in pounds and pence, correct to the nearest penny.

Find the total cost of Option 1 (air tickets + car hire + petrol), and record it to the nearest hundred pounds.

Option 2

Look in the distance chart for Great Britain for the distance from Bournemouth to Fishguard.

Calculate the total miles for the return journey in Great Britain.

Use the figure for average speed, given on the information sheet, to estimate how long the journey in each direction takes to the nearest hour. (time = miles ÷ speed)

Work out the best ferry time for the Smiths to travel out to Ireland and what time they arrive.

Work out which ferry the Smiths must catch to return to Great Britain in time to drive home.

How many days will they have in Ireland?

Look at the map of Ireland to work out a route between the towns that the Smiths want to visit, starting and ending in Rosslare, (near to Wexford).

Look in the distance chart for Ireland to find the distances between the towns.

Calculate the total miles in Ireland.

Work out the grand total for the miles in both countries.

Calculate the cost of all the petrol used for the total distance. Record the answer in pounds and pence, correct to the nearest penny.

Look in the price tables to find the cost of the ferry for one car and a family of four.

Calculate the total cost of Option 2 (petrol + ferry), and record it to the nearest hundred pounds.

****

Check all the information you used.

Check that you did the right calculations to get the results you needed

 You can now ask your Assessor to check this stage for achievement of N1.1

Check all your calculations to see that they make sense.
For your portfolio, write a comment on how you can be sure that the results make sense.

 You can now get your Assessor to check this stage for achievement of N1.2, before you go on to the next stage.

****

Presenting Findings

Compare the cost of Option 1 with the cost of Option 2.
Decide which is the cheapest option.

Compare the times involved in each option. Which option suits the Smiths best for time?

Write a brief letter to the Smiths telling them what you have found out and recommending an option. Include a chart or a diagram, which will help them to see what you mean.

 When you are satisfied that your presentation of findings meets the purpose, ask your Assessor to check your work for achievement of N.1.3

EVIDENCE CHECK

Option 1 Calculations

Cost of air ticket £150

$40\% \text{ of } £150 = \dfrac{40}{100} \times 150 = £60$

£150 – £60 = £90 = price of one ticket

For family of four, tickets cost 90 × 4 = 360   £360.00

Flight Times

Arrive in Dublin at approx. 3.00 pm on first day.

Check in at Dublin to return at 11.00 am on last day.

The Smiths can have 7 days in Ireland, so they need the car for *one week*.

Cost of hiring the cheapest family car for one week is £105.00

Holiday Route in Ireland

Dublin – Cork – Killarney – Dublin

Distances

Dublin – Cork	159
Cork – Killarney	55
Killarney – Dublin	188
Total miles in Ireland	402 miles

Cost of fuel = 10p per mile = 402 × 10 = 4020 p = £40.20

Total cost of option 1 = tickets + car hire + fuel
    £360.00 + £105.00 + £40.20 = £505.20

Total cost of option 1 = £500 to the nearest hundred pounds.

Option 2 Calculations

Distance Bournemouth – Fishguard = 233 miles
Total miles in Great Britain = 233 × 2 = <u>466 miles</u>

At an average speed of 40 mph,
time = miles ÷ speed
time = 233 ÷ 40 = 5.825 hours
To the nearest hour the journey takes 6 hours each way.

The Smiths could leave home early and catch the 13.30 ferry
that gets to Rosslare at 15.30

For their return journey they must catch the 0800 ferry from
Rosslare. The ferry gets to Fishguard at 10.00 am. Then they
will take 6 hours to get home, so they will be home before dark.

The Smiths will have to spend the whole of their last day
travelling home. This means they will only have 6 days in
Ireland if they go by ferry.

Route in Ireland

Wexford – Dublin – Killarney – Cork – Wexford

Distances in miles

Wexford – Dublin	82
Dublin – Killarney	188
Killarney – Cork	55
Cork – Wexford	116
Rosslare – Wexford	10 (estimate)

Total miles in Ireland = <u>451 miles</u>

Grand total of miles in both countries = 466 + 451 = <u>917 miles</u>

Cost of fuel = 10 p per mile = 917 × 10 = 9170 p = <u>£91.70</u>

Cost of ferry

The cheapest ferry price for 7 days is the midweek return
family rate of <u>£289.00</u>

Total cost of option 2 = fuel + ferry
Total cost of option 2 = £91.70 + £289.00 = £380.70

**Total cost of option 2 = £400 to the nearest hundred
pounds**

Check that results make sense:

Looking at the map, the distance from Bournemouth to Fishguard is slightly more than the distance from Dublin to Killarney, so the numbers look about right.

Some of the numbers were looked up twice, so that was a check that they were correct.

The air fare is per person and everyone is charged the same. The ferry price is mostly for the car and driver. Extra people do not cost much extra each; so you would expect the air fare to cost more than the ferry.

The time in Ireland is bound to be less if they travel by ferry because so much time is lost getting to and from the ferry. The journey is 9 hours each way by ferry and 2 hours each way by air.

Comparing option 1 with option 2:

Price: Option 2, travel by ferry, is cheaper. £400 compared with £500

Time: Option 1 gives more time in Ireland. 7 days compared with 6 days.

Presentation of findings

Dear Smith Family,

I have investigated the options for your holiday in Ireland. My results are in the table that follows.

Option	Method of Travel	Approximate Price for Family
1	fly + hire a car	£500
2	use own car and ferry	£400

The price for option 1 is based on
- booking early to get the maximum reduction of 40% on the price of the air tickets
- hiring the smallest family car

If you book later, or want a bigger car, this option will cost more.

Here is a diagram showing the route I used for my calculations for this option.

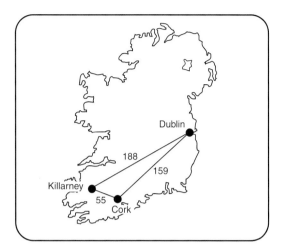

The price for option 2 is based on crossing on the mid-week ferry. If you go at a weekend it will cost more.

The other disadvantage of option 2 is that you will lose one day from your holiday, because you have a long journey home from the ferry.

On the top of the next page is a diagram showing you the route I used for my calculations.

I hope you find this information helpful. You can see my calculations on the attached pages.

Yours sincerely.

Help desk

Charts and Diagrams

All kinds of illustrations can be called "charts" or "diagrams".

They are all ways of showing numerical information.

You must use two different **types** of illustration in your portfolio.

The most important thing is that each type of illustration that you use must be the **most suitable** way of showing the information that you want to give.

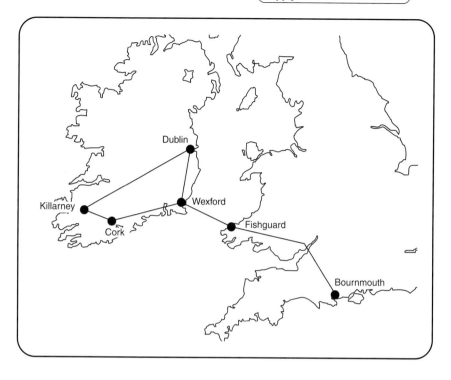

Check the Spec

This activity did not cover handling statistics. To achieve
the whole unit, another activity, which involves statistics
is needed.

Help desk

You must do at least one calculation from each of the categories,
(a), (b) and (c).

In the activity above there was a calculation on proportions
(percentages). There does not have to be a calculation on scales
as well.

A calculation on scales could come into another activity, but it
does not have to.

Specification	Achieved	How achieved
N1.1 Interpret straightforward information from two different sources, including a table, chart, diagram, or line graph	✓	Information came from distance charts, price tables and Activity Brief
• Obtain information to meet the purpose of the task	✓	The information from the charts and tables could be used to do calculations and tell the Smiths what they wanted to know
• Identify suitable calculations to get the results you need	✓	The calculations gave the information that was wanted
N1.2 Carry out straightforward calculations to do with • amounts and sizes • scales and proportions • handling statistics	✓ ✓ ✗	Prices and distances Percentage calculation Not required for this activity
• Carry out calculations to the level of accuracy given	✓	Rounding prices – see calculations pages
• Check your results make sense	✓	See calculations pages
N1.3 Interpret the results of calculations and present findings, using: one chart one diagram	✓ ✗ ✓	Comparing option 1 with option 2 – see the letter to the Smiths, with the table and diagrams (route maps).
• Choose suitable ways to present findings • Present findings clearly	✓ ✓	Prices set out clearly in a table Route maps to show distances
• Describe how the results of your calculations meet the purpose of the task	✓	The letter tells the Smiths what they want to know They can make a decision by using the information in the letter The diagrams help them to see how the figures were worked out.

Further Activities for Portfolio Assignments

Carry out an opinion poll on a topic of current interest.

Carry out a census to find what proportions of people have particular lifestyles.

Carry out a consumer survey to find out about people's buying habits.

Produce plans for giving your home or workplace or garden a make-over. Find out the costs of new things and draw some diagrams to show how it all fits together.

Work out the costs, and find the 'break-even point' for:
- Running a small business
- Putting on an entertainment
- Running a visit to a place of interest
- Taking a regular club slot at a leisure centre, or arts centre or sports centre.

There are various ways to show your findings in charts and diagrams.

Help desk

All your Activities should involve work for **N1.1, N1.2 and N1.3**
Each component (or stage) can be **assessed** separately.
You can **achieve** each component independently of the others.

Putting your Portfolio together for Part B

Your portfolio is the evidence that shows you can do all of the things listed in Part B.

Before you put work into your portfolio it must be assessed. You need to agree to a procedure for assessment with your assessors, but the procedure will be something like this.

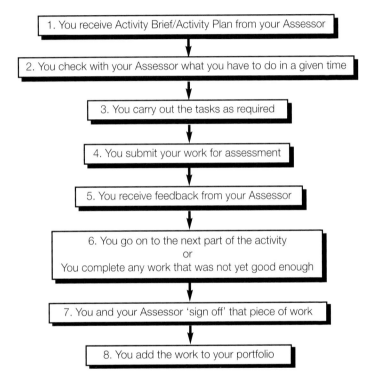

1. You receive Activity Brief/Activity Plan from your Assessor

2. You check with your Assessor what you have to do in a given time

3. You carry out the tasks as required

4. You submit your work for assessment

5. You receive feedback from your Assessor

6. You go on to the next part of the activity
   or
   You complete any work that was not yet good enough

7. You and your Assessor 'sign off' that piece of work

8. You add the work to your portfolio

Help desk

Your Assessor can tell you **what** to do.
It is up to you to make sure you do it well enough to **achieve** the level.

Application of Number Level 2: Developing the Skills and Knowledge in Part A

Introduction

Part A describes the skills and knowledge that you need to **learn** for the Application of Number qualification at Level 2. You may have to pass a test as part of your qualification.

At Level 2 you must be able to carry through a *substantial activity* that requires you to:

- *Select information and methods* to get the results you need
- Carry out calculations *involving two or more steps and numbers of any size, including use of formulae, and check your methods and levels of accuracy*
- *Select ways of presenting your findings, including use of a graph, describe methods and explain results*

You develop these skills in Part A and apply them in Part B. In Part B you are expected to carry out *at least one substantial activity that can be broken down into tasks for each component of the unit*

- All calculations must be clearly set in context
- All calculations must be clearly set out with evidence of checking
- You must show that you are clear about the purpose of your task

We are going to look at Part A first.

Understanding Part A

Read through the list of "bullet points" in Part A of the specification. Probably there will be some things you think you can do, and some things you are not sure about. If you have not already done it, turn to the **self-assessment check-list** on page 11 and complete it. To pass the test you need to be sure that you can do all of Part A.

The test is a multiple-choice paper. It has 40 questions. The questions are designed to find out if you can do all of the things listed in Part A.

To get the right answers you have to

• Read and understand the information in the question. Sometimes numbers are in the text of the question. Sometimes there is a table, chart, graph or diagram that you must interpret

• Do calculations with the numbers you obtained

• Select the correct answer from the options, A, B, C or D

You are not allowed to use a calculator in this test. You can jot down the information, but then you must do the calculation in your head.

The next section helps you to check how much you already know, and what you need to practice.

As you go through this section, work out the answers to the questions **without** a calculator, and **then** check your answers. Make sure that you understand **how** the correct answer is worked out.

Numbers and Calculations

Rules for calculating

At Level 2 you are expected to be able to carry out calculations that involve 2 or more steps. To do this you need to be sure that you understand the rules for calculating.

It is easy to write down calculations which you understand in one way but which other people may read differently.

For instance, what is the correct answer to $2 + 3 \times 4 - 3$?

You can get 5, 11, or 17 depending how you read it.

The 'operations' of $+$, $-$, $\times$, and $\div$ must be carried out in the correct order

Operations Help desk

Order of operations

Rule 1 Work from left to right

Rule 2 Do $\times$ and $\div$ before doing $+$ and $-$.

Apply the rules to $2 + 3 \times 4 - 3$

$$3 \times 4 = \quad 12$$
$$2 + 12 - 3 = \quad 11$$

Try these questions
1   (i) $12 - 4 \times 2$,   [4]
    (ii) $12 \times 2 - 4 \div 2$,   [22]
    (iii) $12 \div 3 + 4 \times 4 - 1$ [19]

> If your answers were incorrect, check back with the rules above and make sure you carried out the operations in the correct order

Using brackets

Some calculations will involve brackets.
Brackets make clear which parts of a calculation to do first.

Operations Help desk

Rule 1 Work from left to right
Rule 2 Work out $\times$ and $\div$ before working out $+$ and $-$
Rule 3 Do the calculations inside brackets first

Remember: the $\times$ signs outside brackets are often omitted
2   What do the following expressions mean?
    $3 + 4 \times 5$ means 'add 3 to the result of 4×5' $= 3 + 20 = 23$.
    $(3 + 4) \times 5$ means 'multiply the result of $3 + 4$ by 5' $= 7 \times 5 = 35$
    $(3 + 4)5$ means 'multiply the result of $3 + 4$ by 5' $= 7 \times 5 = 35$

3   Work out these calculations without using your calculator.
    (i) $4 \times 7 + 5 \times 3$,   [43]
    (ii) $4(7 + 5)3$,   [144]
    (iii) $\dfrac{4(5 - 2.6)}{8}$ [1.2]
    (vi) $4 \times 5 - \dfrac{2.6}{8}$ [19.675]

Directed numbers

All numbers have direction! They can move in a positive
direction $\rightarrow$ getting bigger, or in a negative direction $\leftarrow$
getting smaller. This can be seen clearly using a number line.

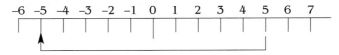

Example:
Work out   5 – 10
Find 5 on the number line,
Move 10 places to the left, in the **negative** direction
The answer is –5
5 – 10 = –5   –5 is a **negative number**

Negative numbers occur in practical situations. When a bank account 'goes into the red' it means that the balance is negative.

Example
John had £42.30 in his bank account
He spent £50.
How much was in his account?
£42.30 – £50 = –£7.70

In winter months the temperature often drops below freezing point. Negative numbers are used to show temperatures below freezing.

Example
During the day the temperature was 4°C.
At night the temperature dropped by 8°C.
What was the night-time temperature?
Night-time temperature = (4 – 8)°C = –4°C

Try these questions
1   The temperature of the fridge compartment of a fridge freezer is set at 8°C. The freezer compartment is set at – 12° C. What is the difference in these settings? [20°C]
2   Temperatures in New York City

	Jan	Feb	Mar	Apr	May	Jun	Jul	Aug	Sep	Oct	Nov	Dec
Ave Daily High C	3	3	7	14	20	25	28	27	26	21	11	5
Ave Daily Low C	–4	–4	–1	6	12	16	19	19	16	9	3	2

What was the difference between the high and low average temperatures in Jan?
[3 – –4 = 7]

Powers

Look at these quick ways of showing repeated multiplying.

You can write $\quad 4 \times 4 \qquad$ (16) $\quad$ as 4^2

You can write $\quad 5 \times 5 \times 5 \quad$ (125) $\quad$ as 5^3

The small raised number (called the 'power' or 'index') tells you how many times the main number is multiplied by itself.

Write down what the following multiplyings are, then work them out.

(i) $\quad 4^3 \qquad [4 \times 4 \times 4 = 64]$

(ii) $\quad 3^5 \qquad [3 \times 3 \times 3 \times 3 \times 3 = 243]$

(iii) $\quad 2^6 \qquad [2 \times 2 \times 2 \times 2 \times 2 \times 2 = 64]$

(iv) $\quad 7^3 \qquad [7 \times 7 \times 7 = 343]$

(v) $\quad 10^5 \qquad [10 \times 10 \times 10 \times 10 \times 10 = 100\,000]$

(vi) $\quad 10^7 \qquad [10 \times 10 \times 10 \times 10 \times 10 \times 10 \times 10 = 10\,000\,000]$

Tell the story of your calculation

At Level 2 you must show your calculations clearly and fully. When you are doing calculations it's important to be able to check what you have done for mistakes. Later you may well have to convince other people that your calculations are correct – this means explaining them so that other people can understand.

Get in the habit of writing down what your figures mean instead of just writing the figures. This means using words as well as numbers!

Example:

A milkman delivers 2 bottles of milk each day of the week to a house. The milk costs 32p per bottle.

(i) What would be the weekly bill?

(ii) What would be the cost of milk for one year?

Solution

(i) Cost per bottle $\qquad\qquad\qquad\qquad$ 32p

Daily cost $\qquad\qquad\qquad 2 \times 32$ p $\quad = 64$p

Weekly cost $\qquad\qquad\quad 7 \times 64$p $\quad = 448$p

Convert the total cost to pounds

Weekly cost in pounds $\qquad 448 \div 100 \quad = $ **£4.48**

(ii) There are 52 weeks in a year

Cost for 52 weeks $\qquad\qquad$ £4.48 $\times 52 = $ **£232.96**

This is much easier to understand than:

32

64

448

23296

Checking calculations

At Level 2 you must show how you check the results of your calculations.

Approximations can be used to check your results

Example :

Calculation: $20.5 \times 251.7 = 5159.85$

Approximation: $20 \times 250 = 5000$

Example:

The cost of framing a picture is £5.99 plus £8.99 per metre of framing needed.

The length needed is 1.9 m

(i) Work out the cost

(ii) Use approximation to check your answer

Calculation Help desk

(i) Cost $=£5.99 + 1.9 \times £8.99 = £23.071 = £23.07$ (to 2 d.p.)

(ii) Approximation $= £6 + £(9 \times 2) = £24.00$

Use approximation to check these answers:

1. $36.98 \,\text{cm} + (8.2 \times 13.7) \,\text{cm} = 149.32 \,\text{cm}$

    $[37 + (8 \times 14) = 149]$

2. $£34.78 + £51.02 - £64.97 = £20.63$

    $[35 + 51 - 65 = 21]$

Reverse calculations

Calculation		Check	
Addition	$+$	Subtraction	$-$
Subtraction	$-$	Addition	$+$
Multiplication	$\times$	Division	$\div$
Division	$\div$	Multiplication	$\times$

Addition
345.23 + 642.78 = 988.01
Check by subtraction
988.01 − 642.78 = 345.23

Subtraction
45.357 − 51.904 = −6.547
Check by addition
−6.547 + 51.904 = 45.357

Multiplication
8.2 × 5.6 = 45.92
Check by division
45.92 ÷ 5.6 = 8.2

Division
428.0164 ÷ 19.88 = 21.53
Check by multiplication
21.53 × 19.88 = 428.0164

Use reverse calculations to check these answers
1   113.54 + 66.73 + 84.52 = 264.79
     [subtraction 264.79 − 84.52 − 66.73]
2   −46.814 − 12.125 − 21.873 = −80.812
     [addition −80.812 + 21.873 + 12.125]
3   437.56 × 34.002 = 14877.91512
     [division 14877.91512 ÷ 34.002]
4   89.656 ÷ 22.414 = 4
     [multiplication 22.414 × 4]

Conversions

You need to know how to:
- Convert between currencies
- Convert from metric to imperial and imperial to metric
- Convert within the metric system

Currency conversion

Remember: the level of accuracy for currency is **2** decimal places

Examples

1   The exchange rate is 1.41 US dollars ($) to the pound.
        Change £500 into $.
        £1 is 1.41$
        £500 is 500 × 1.41$= 705$.
2   The exchange rate is £0.62 to one euro (€1)
        Change €69 to pounds
        €1 is £0.62
        69 × 0.62 = £42.78
3   The exchange rate is 14.95 (Swedish) kroner to the pound.
        Change 2000 kroner into pounds.

Calculation

14.95 kroners are £1

1 ÷ 14.95 = 0.066 889

1 kroner = £0.066 889

2000 kroner = £0.066 889 × 2000

= . £133.778 = £133.78 to 2 d.p.

Check

2000 ÷ 14.95 = £133.7792642

= £133.78 to 2 d.p.

Question 1

The exchange rate is 1.61 euros to the pound.

Change £375 into euros.

Question 2

The exchange rate is 532 (Greek) drachma to a pound.

Change 30 000 drachma into pounds.

Answers:

1 603.75 euros

2 £56.39

Help desk

1 £1 is 1.61 euros.
   £375 is 375 × 1.61 euros = 603.75 euros.   **Check:** 603.75 ÷ 1.61 = 375

2 532 drachma is £1.
   30 000 drachma is 30 000 ÷ 532 = £56.39.   **Check** 56.39 × 532 = 30 000

Converting between imperial and metric measurements

Imperial/metric conversions

Length

Imperial	→	metric	Metric	→	imperial
1 mile	→	1.6 km	1 km	→	$\frac{5}{8}$ mile
1 foot	→	30 cm	1 m	→	39 inches
1 inch	→	2.5 cm	1 mm	→	0.039 in

Weight

Imperial	→	metric	Metric	→	imperial
1 ton	→	1.02 t	1 t	→	0.984 ton
1 Pound(lb)	→	450 g	1 kg	→	2.2 lbs
1 oz	→	28 g	1 g	→	0.353 oz

Capacity

Imperial	$\rightarrow$	metric	Metric	$\rightarrow$	imperial
1 gal	$\rightarrow$	4.5 l	1 l	$\rightarrow$	0.22 gal
1 pt	$\rightarrow$	0.57 l	1 l	$\rightarrow$	1.76 pt
1 fl.oz	$\rightarrow$	28.4 ml	1 ml	$\rightarrow$	0.0352 fl.oz

Remember: each of these conversions is approximate.

Example

Change 220 miles into kilometres.

1 mile is 1.6 km so 220 miles is 220×1.6 km = 352 km

Example

Change 5 kg into pounds.

1 kg is 2.2 lbs so 5 kg is 5×2.2 lbs = 11 lbs

■ NOTE: Check that your answers are sensible.

In the first example a kilometre is less than a mile so we expect more kilometres. (352 is more.)

In the second example a pound is less than a kilogram so we expect more pounds. (11 is more.)

Question 1   Change 16 inches into centimetres.

Question 2   Change 11 gallons into litres.

Question 3   Change 5 ounces into grams.

Answers:
1. 40 cm
2. 49.5 litres
3. 140 grams

Calculation Help desk

Convert		Check	
1. 16 ins = 16×2.5 cm	= 40 cm.	$40 \div 2.5$	= 16
2. 11 galls = 11×4.5 litres	= 49.5 litres.	$49.5 \div 4.5$	= 11
3. 5 oz = 5×28 ounces	= 140 grams	$140 \div 28$	= 5

Converting within the metric system

Metric Measures

Length

Kilometre (km)			1 km	=	1000 m
Metre (m)	1000 m	= 1 km	1 m	=	100 cm
Centimetre	100 cm	= 1 m	1 cm	=	10 mm
Millimetre	10 mm	= 1 cm			

Weight

Tonne (t)			1 t	=	1000 kg
Kilogram (kg)	1000 kg	= 1 t	1 kg	=	1000 g
Gram (g)	1000 g	= 1 kg	1 g	=	1000 mg
Milligram	1000 mg	= 1 g			

Capacity

Litre (l)			1 l	=	100 cl
Centilitre (cl)	100 cl	= 1 litre	1 cl	=	10 ml
Millilitre (ml)	10 ml	= 1 cl			
Cubic metre (m^3)	1000 l	= 1 m^3	1 m^3	=	1000 l
Cubic centimetre (cm^3)	1 ml	= 1 cm^3	1 cm^3	=	1 ml

Examples:

Convert the following measurements:

1. 7685 m to km
2. 4.5 cm to mm
3. 2352 kg to t
4. 1.5 g to mg
5. 345 cl to l
6. 7523 l to m^3

Conversion Help desk

Convert			Check	
1. 7685 m = 7685 ÷ 1000 km	= 7.685 km		7.685 × 1000	= 7685
2. 4.5 cm = 4.5 × 10 mm	= 45 mm		45 ÷ 10	= 4.5
3. 2351 kg = 2351 ÷ 1000 t	= 2 351 000 kg		2.351 × 1000	= 2351
4. 1.5 g = 1.5 × 1000 mg	= 1500 mg		1500 ÷ 1000	= 1.5
5. 345 cl = 345 ÷ 100 l	= 3.45 l		3.45 × 100	= 345
6. 7523 l = 7523 ÷ 1000 m^3	= 7.523 m^3		7.523 × 1000	= 7523

Fractions, decimals and percentages

You need to know how to convert:

- Fractions to percentages and percentages to fractions
- Decimals to percentages and percentages to decimals.
- Fractions to decimals and decimals to fractions.

Converting between fractions and percentages

Signs such as:

Sale!
25% off

Sale!
¼ off

are examples which show that you need to know how to convert between fractions and percentages.

25% means 25 out of every 100. So $25\% = \dfrac{25}{100}$.

$\dfrac{25}{100}$ can be cancelled to $\dfrac{1}{4}$ so $25\% = \dfrac{1}{4}$.

> **Definition:** a fraction that cannot be cancelled down is said to be in its lowest terms or simplest form

Example

Change 40% to a fraction in its lowest terms.

$40\% = \dfrac{40}{100}$ and $\dfrac{40}{100} = \dfrac{2}{5}$. So $40\% = \dfrac{2}{5}$.

Help desk

The fractions $\dfrac{40}{100}$, $\dfrac{20}{50}$, $\dfrac{10}{25}$ and $\dfrac{2}{5}$ all have the same value.

Only $\dfrac{2}{5}$ is in its lowest terms. It cannot be cancelled down any more.

A fraction is changed to a percentage by multiplying by 100.

Example

Change $\dfrac{3}{4}$ to a percentage.

$\dfrac{3}{4} \times 100 = 75$. So $\dfrac{3}{4}$ is 75%.

Question 1

Change the following to fractions in their lowest terms.

(i) 80%   (ii) 65%

Question 2

Change the following to percentages.

(i) 1/5   (ii) 21/25

Question 3

3 of the 20 residents in a care home are male.

What percentage are male?

Question 4

In a school 12 boys out of 340 are colour-blind.

What percentage are colour-blind?

Answers:

1 (i) 4/5   (ii) 13/20
2 (i) 20% (ii) 84%
3   15%
4 3.5% approx.

Help desk

1 (i) $80\% = \dfrac{80}{100} = \dfrac{4}{5}$ (dividing by 20) $80\% = \dfrac{4}{5}$.

   (ii) $65\% = \dfrac{65}{100} = \dfrac{13}{20}$ (dividing by 5) $65\% = \dfrac{13}{20}$.

2 (i) $\dfrac{1}{5} \times 100 = 20$ $\dfrac{1}{5} = 20\%$.

   (ii) $\dfrac{21}{25} \times 100 = 84$ $\dfrac{21}{25} = 84\%$.

3 $\dfrac{3}{20} \times 100 = 15$ 15% are male.

4 $\dfrac{12}{340} \times 100 = 3.529\ldots$ 3.5% (to 1d.p.) are colour-blind.

Converting between decimals and percentages

Decimal to percentage

Change 0.6 to a percentage.

$0.6 \times 100 = 60$, so 0.6 is 60%.

Percentage to decimal

Change 45% to a decimal.

$$45\% = \dfrac{45}{100} = 0.45$$

Question 1

Change the following to percentages.

(i) 0.72   (ii) 0.375

Question 2

Change the following to decimals.

(i) 54%   (ii) 12.5%

Answers:

1 (i) 72%   (ii) 37.5%

2 (i) 0.54   (ii) 0.125

Help desk

1 (i) $0.72 \times 100 = 72$ $0.72 = 72\%$

(ii) $0.375 \times 100 = 37.5$ $0.375 = 37.5\%$

2 (i) $54\% = \dfrac{54}{100} = 0.54$ $54\% = 0.54$

(ii) $12.5\% = \dfrac{12.5}{100} = 0.125$ $12.5\% = 0.125$

Converting between fractions and decimals

A fraction is changed to a decimal by division.

Help desk

Fraction to decimal

Change $\dfrac{3}{4}$ to a decimal.

$\dfrac{3}{4} = 3 \div 4$ and $4 \overline{)3.00}$ giving 0.75 so $\dfrac{3}{4} = 0.75$

Decimal to fraction

A decimal is changed to a fraction by considering place value.

Change 0.75 to a fraction.

0.75 is 75 hundredths and $\dfrac{75}{100} = \dfrac{15}{20} = \dfrac{3}{4}$

Question 1

Change these fractions to decimals.

(i) $\dfrac{2}{5}$ (ii) $\dfrac{3}{20}$

Question 2

Change these decimals to fractions.

(i) 0.9 (ii) 0.71

Answers:

1 (i) 0.4 (ii) 0.15

2 (i) $\dfrac{9}{10}$ (ii) $\dfrac{71}{100}$

Help desk

1 (i) $\dfrac{2}{5} = 2 \div 5$ and $5\overline{)2.0}^{\,0.4}$ so $\dfrac{2}{5} = 0.4$

(ii) $\dfrac{3}{20} = 3 \div 20$ and $20\overline{)3.00}^{\,0.15}$ so $\dfrac{3}{20} = 0.15$

2 (i) 0.9 is nine tenths so $0.9 = \dfrac{9}{10}$

(ii) 0.71 is seventy one hundredths so $0.71 = \dfrac{71}{100}$

Understanding proportions and ratios

Proportions are often given as percentages or fractions.

Example

A holiday costs £650.

The price is increased by 8%.

The increase is $\dfrac{8}{100} \times £650 = £52$

The new price is £650 + £52 = £702.

■ NOTE: the new price can also be found by multiplying £650 by 1.08.

Question 1

Emily buys a new coat priced at £80.

She gets a 20% discount.

How much does Emily pay?

Question 2

A computer costs £799.95 excluding VAT.

VAT is added at 17.5%.

What is the total cost?

Answers:

1 £64

2 £939.94

Calculation Help desk

1 20% of £80 = $\dfrac{20}{100}$ × £80 = £16. Emily pays £80 – £16 =£64.

2 17.5% of £799.95 = $\dfrac{17.5}{100}$ × 799.95 = 139.99125. VAT is £139.99 (nearest penny).

Total cost is £799.95 + £139.99 = £939.94

Example

A pair of trainers is priced at £69 before a sale.

In a sale prices are reduced by one third.

The amount saved is $\dfrac{1}{3}$ of £69 = $\dfrac{1}{3}$ × £69 = $\dfrac{£69}{3}$ = £23

The sale price is £69 – £23 = £46.

■ NOTE: the sale price can also be found by finding two thirds of £69.

Question 1

Kelly's car insurance last year was £348.

This year it is increased by a quarter.

How much is her insurance this year?

Question 2

Last year a company made a profit of £35 million.

It is expected that the profit this year will be down by a fifth.

What is the expected profit this year?

Answers:

1 £435

2 £28 million

Calculation Help desk

Calculation	Check
1. $\frac{1}{4}$ of £348 = £348 ÷ 4 = £87	87 × 4 = 348
£348 + £87 = £435	435 − 87 = 348
2. $\frac{1}{5}$ of £35 million = £35 ÷ 5 = £7 million	7 × 5 = 35
£35 million − £7 million = £28 million	28 + 7 = 35

Ratio

Proportions can also be expressed using ratio.

Example

Phil and Amy share a profit of £12 000 in the ratio 3:2.

This means that for every £3 Phil receives, Amy receives £2.

For every 3 parts that Phil receives Amy receives 2 parts

so there are 3 + 2 = 5 parts in total.

5 parts is £12 000.

1 part is $\dfrac{£12000}{5}$ = £2400

Phil receives 2 × £2400 = £4800.

Amy receives 3 × £2400 = £7200.

You can see that sharing in a given ratio could also be described in terms of fractions or percentages.

Phil receives 3 parts out of 5 so Phil gets $\frac{3}{5}$ (60%) of the profit and Amy gets $\frac{2}{5}$ (40%).

Question 1

Raj makes orange paint by mixing red and yellow in the ratio 1:2.

He wants 24 litres of orange paint.

How much of each colour does he need?

Question 2

One unit of cleaning fluid is mixed with 20 units of water.

How many units of cleaning fluid are mixed with

(a) 60 units of water   (b) 150 units of water?

Calculation Help desk

1. Total number of parts = 1 + 2 = 3
   One part = 24 ÷ 3 = 8 litres
   Red = 1 × 8 litres = 8 litres
   Yellow = 2 × 8 litres = 16 litres
2. (a) ratio of 1:20
       60 ÷ 20 = 3
       1 × 3 = 3 units of cleaning fluid
   (b) ratio of 1:20
       150 ÷ 20 = 7.5
       1 × 7.5 = 7.5
       7.5 units of cleaning fluid

Work out dimensions from scale drawings

Here is a scale drawing showing the design for Hana's new office.

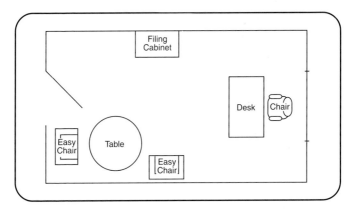

The scale is 1:50. This means that one unit on the drawing represents fifty units in real life.

For example 1 cm on the drawing represents 50 cm in real life.

Example

On the drawing the length of the office is 9 cm.

The actual length is 50 × 9 cm = 450 cm or 4.5 m.

■ NOTE: To convert centimetres to metres multiply by 100.

Example

On the drawing the diameter of the table is 1.8 cm.

The actual diameter is 50×1.8 cm = 90 cm.

■ NOTE: In each example you should think about the answer and check that it is sensible.

By doing the wrong calculation (e.g. dividing by 50 instead of multiplying) you might get

an office of length $9 \div 50 = 0.18$ cm, or

a table of diameter $1.8 \div 50 = 0.036$ cm

In each case you should realise that the answer isn't sensible.

Question 1

What is the width of the office?

Question 2

What is the width of the desk?

Answers:

1   2.5 m

2   0.6 m

Calculation *Help desk*

Calculation helpdesk

1. Scale measurement = 5 cm

   $5 \times 50 = 250$ cm

   $250 \div 100 = 2.5$ m

2. Scale measurement = 1.2 cm

   $1.2 \times 50 = 60$ cm

   $60 \div 100 = 0.6$ m

Perimeter, area and volume

Definition *Help desk*

Perimeter:	the distance all around a shape. Add together the lengths of all the sides
Area:	the space inside the perimeter – measured in square units
Volume:	the area of the cross-section x height – measured in cubic units

Find the perimeter and area of this rectangular hall

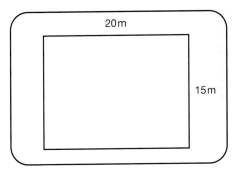

20 m

15 m

Perimeter = 2 × (length + width)

Area = length × width

Perimeter = 2 × (20 + 15) = 2 × 35 = 70.   Area = 20 × 15 = 300
The perimeter is 70 m   The area is 300 m².

Remember: Area is found by multiplying two lengths so it is
measured in **square units**. In this case the units are
square metres and they are written **m²**.

Example

The hall described in the last example is 4 m high.

Find its volume.

Volume = 20 × 15 × 4 = 1200.   Volume = length × width × height

The volume is 1200 m³.

Remember: Volume is found by multiplying three lengths so it is
measured in **cubic units**. In this case the units are
cubic metres and they are written **m³**.

Question 1

For this L-shaped room find (i) the area and (ii) the perimeter.

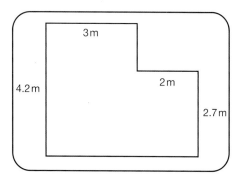

Question 2

A metal plate is 10 cm long, 4 cm wide and 2 mm thick.
Find the volume of the plate.

Here is a familiar box shape (you may find it called a cuboid).
All the measurements are in the same 'family' of units.

Question 3

A box is 40 cm by 25 cm by 26 cm.

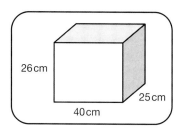

(i) How many packets 10 cm by 5 cm by 2 cm fit into it?
(ii) How many packets 20 cm by 12.5 cm by 8 cm fit into it?

Answers

1 (i) Perimeter = 18.4 m   (ii) Area = 18 m²
2 8 cm³ (or 8000 mm³)
3 (i) 260   (ii) 12

Calculation Help desk

1 (i) The missing lengths are 3 m + 2 m = 5 m and 4.2 m − 2.7 m = 1.5 m
   Perimeter is 3 + 1.5 + 2 + 2.7 + 5 + 4.2 = 18.4
   Perimeter = 18.4 m

 (ii) The area is the sum of two rectangles.
   Area is (4.2 × 3) + (2.7 × 2) = 12.6 + 5.4 = 18
   Area = 18 m².

2 Notice there is a mix of units. You must choose cm or mm.
  Using cm the volume is 10 × 4 × 0.2 = 8
  Volume = 8 cm³.
  Using mm the volume is 100 × 40 × 2 = 8000
  Volume = 8000 mm³.

3 Match dimensions and fit as many as you can in each direction.

 (i)

	Length	Width	Height
Box	40	25	26
Packet	10	5	2
	40 ÷ 10 = 4	25 ÷ 5 = 5	26 ÷ 2 = 13

   The total number of packets is 4 × 5 × 13 = 260.

 (ii)

	Length	Width	Height
Box	40	25	26
Packet	20	12.5	8
	40 ÷ 20 = 2	25 ÷ 12.5 = 2	26 ÷ 8 = 3.25 (i.e. only 3 complete packets)

   The total number of packets is 2 × 2 × 3 = 12

You could place these packets so that width = 8 cm and height = 12.5 cm
This would also give 12 packets. (2 × 3 × 2)

Circles

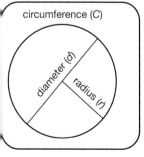

circumference (C)

diameter (d)

radius (r)

Circle Definitions
Centre: the point at the middle of a circle
Circumference: the distance all around a circle
Radius: the distance from the centre to any
point on the circumference
Diameter: any straight line joining two points on
the circumference and passing through the centre

Circle helpdesk

The formulas connected with circles involve π (pi)

Circumference of a circle

$$C = \pi \times d = \pi d$$
$$C = 2 \times \pi \times r = 2\pi r$$

Area of a circle

$$A = \pi \times \frac{d \times d}{4} = \frac{\pi d^2}{4}$$
$$A = \pi \times r \times r = \pi r^2$$

Remember:

The diameter is twice as long as the radius

The radius is equal to the diameter divided by 2

Example

A circle has radius 5 m, calculate:

(i) the circumference

(ii) the area.

Calculation Help desk

(i) The circumference is given by $C = 2\pi r = 2\pi \times 5 = 31.4$.

　　　Circumference is 31.4 m.

Check: by using $\pi = 3$　　　$2 \times 3 \times 5 = 30\,m$

■ NOTE: if your calculator has a π button then you should use this rather than key in the approximation 3.14

(ii) The area is given by $A = \pi r^2 = \pi \times 5 \times 5 = 78.5$.

　　　Area is 78.5 m².

Check: by using $\pi = 3$　　　$3 \times 5 \times 5 = 75\,m^2$

Question 1

The radius of a circle is 8.2 cm.

Work out (i) the circumference and (ii) the area.

Question 2

The diameter of a circle is 46 cm.

Work out (i) the circumference and (ii) the area.

Answers:
■ NOTE: These answers are all approximate.

1 (i) 51.5 cm   (ii) 211.2 cm^2
2 (i) 144.5 cm   (ii) 1662 cm^2

Calculation Help desk

1 (i) $C = 2\pi r$ so $C = 2\pi \times 8.2 = 51.5$
   (ii) $A = \pi r^2$ so $A = \pi \times 8.2^2 = 211.2$
2 $d = 46$ so $r = \dfrac{46}{2} = 23$
   (i) $C = 2\pi r$ so $C = 2\pi \times 23 = 144.5$
   (ii) $A = \pi r^2$ so $A = \pi \times 23^2 = 1662$

Using formulas (■ NOTE: sometimes we use **formulae**)

Formulas are just shorthand ways of writing down 'recipes' for particular groups of calculations.

In a formula, you use letters to stand for values of what are called variables. Multiplication signs ($\times$) between variables are often omitted in formulas.

Rectangles
Perimeter $P = 2(L + W)$,
Area $A = L \times W$,
Volume $V = L \times W \times H$

Circles
Circumference $C = 2\pi r$,
Area $A = \pi r^2$.

You will meet formulas in many other contexts. For example:

$F = 1.8C + 32$ Converts degrees Celsius into degrees Fahrenheit.

$V = IR$ Uses electrical current and resistance to find potential difference. This is Ohm's Law.

$B = M/h^2$ Uses body mass and height to determine whether a person is under/overweight. This is the formula for Body Mass Index.

$$\boxed{C = F + nU}$$

Uses fixed cost and unit cost to calculate the total cost of producing n items.

Example

Harry wants to print 200 leaflets advertising his disco.

He gets a quote from a printer who gives him the formula

$$\boxed{C = 12 + 0.05 \times n}$$

where C is the total cost in pounds and n is the number of leaflets required.

This means the total cost is £12 plus 5 pence per leaflet.

The cost of 200 leaflets is found by putting $n = 200$ in the formula.

$C = 12 + 0.05 \times 200 = 12 + 10 = 22$

The cost of 200 leaflets is £22.

Question 1

Use the formula $C = 12 + 0.05 \times n$ to work out
(i) how much Harry pays for 500 leaflets.
(ii) how many leaflets Harry gets for £30.

Answers:
1 (i) £37   (ii) 360

Help desk

1 (i) Put $n = 500$ in the formula to get
$C = 12 + 0.05 \times 500 = 12 + 25 = 37$
(ii) Put $C = 30$ in the formula to find n
$30 = 12 + 0.05 \times n$
$30 - 12 = 0.05n$
$18 = 0.05 \times n$
$18 \div 0.05 = n$
$n = 360$

Working with statistics: Mean, median, mode and range

Here are the salaries of the employees of Michael's Marketing Company.

£8000	£8000	£8500	£8500
£8500	£8500	£9000	£9000
£9000	£9500	£10000	£10000
£11000	£11500	£14000	£15000
£18000	£23000	£29000	£40000

You can summarise this data by choosing a typical salary, £xxxxx, and say

"On average an employee earns £xxxxx per year".

However you may choose this typical salary in different ways. You could choose mean, median or mode.

Help desk

The **Mean** $= \dfrac{\text{Sum of the items}}{\text{Number of the items}}$

The **median** is the middle number, or the mean of the two middle numbers, in a list of numbers arranged in order of size.

The **mode** is the number that occurs most often.

Mean

This is found by adding the salaries together and dividing by the number of them.

Adding all the salaries together gives a total of £268000. There are 20 salaries so

$$\text{mean salary} = \frac{£268000}{20} = £13400$$

The mean is the most commonly used average. Each number contributes to it.

Median

There are an even number of employees. (These numbers represent £'000)

8 8 8.5 8.5 8.5 8.5 9 9 9 9.5 ▲10 10 11 11.5 14 15 18 23 29 40

⎵_____⎵ middle ⎵_____⎵
         10   10

As there are 20 numbers, the median is found by working out the mean of the middle two numbers

$$\text{median} = \frac{£9500 + £10000}{2} = £9750$$

Mode

£8500 occurs most often (4 times).

mode = £8500

> Remember: Range = highest number – lowest number

Another way of describing data is by measuring its spread. The simplest measure is the **range**.

Highest salary = £40000
Lowest salary = £8000
    Range   = £40 000 – £8 000 = £32 000

Remember: You must work out the range. Just writing £8000 – £40 000 is not good enough!

Question 1

In Michael's Marketing Company, Michael is the employee who earns £40 000.

Work out the mean salary of the other 19 employees.

Question 2

Find the median of the five managers whose salaries are
£17 000   £19 000   £23 000   £28 000   £40 000

Question 3

Six employees have joined the company in the last year.
Their salaries are
£8000   £8000   £8500   £9000   £10 000   £17 000
Find the mode.

Question 4

Use the data in questions 2 and 3 to find the range of the salaries of

(i) the five managers

(ii) the six employees who joined the company in the last year

Answers:

1 £12 000

2 £23 000

3 £8000

4 (i) £23 000   (ii) £9000

Calculation Help desk

1 The total of the salaries is now £268 000 – £40 000 = £228 000.
There are now 19 salaries.

$$\text{Mean} = \frac{£228000}{19} = £12\,000$$

2 The salaries are in order

£17 000   £19 000   £23 000   £28 000   £40 000

↑

middle

£23 000 is the middle one (there are two each side) so median is £23 000.

3 £8000 is the most common. It occurs twice. The mode is £8000.

4 (i) Range = £40 000 – £17 000 = £23 000.

(ii) Range = £17 000 – £8000 = £9000.

Notice how one large salary (£40 000) has a significant effect on the mean.

When it was included the mean was £13 400 and without it the mean fell to £12 000.

In such circumstances the median may provide a more typical salary.

Comparing results

James and Sophie are comparing the salaries of the employees of two businesses.

Michael's Marketing Company (see previous pages) has 20 employees. Their salaries are

£8000	£8000	£8500	£8500
£8500	£8500	£9000	£9000
£9000	£9500	£10000	£10000
£11000	£11500	£14000	£15000
£18000	£23000	£29000	£40000

The mean salary is £13400, the median £9750 and the mode £8500.

The range of salaries is £32000.

Katy's Catering has 25 employees. Their salaries are:

£10000	£10000	£10200	£10400	£10500
£10500	£10600	£10800	£10800	£10900
£11000	£11200	£11600	£12000	£12500
£12800	£13000	£13300	£13900	£13900
£13900	£18500	£19200	£21000	£30000

The mean salary is £13300, the median is £11600 and the mode is £13900.

The range of salaries is £20000.

Comparing two sets of data is rarely easy! For example in this case you are not told how many hours each employee works or whether the businesses are based in the same town ... .

However you must do your best and support your argument with some figures.

James claims that Michael's Marketing pays better.

He says that:
(i) the mean salary is £100 higher
(ii) 25% of the employees earn £15000 or more whereas only 16% of the employees of Katy's Catering do.

Sophie claims that Katy's Catering pays better.

She says that:
(i) the median salary is £1850 higher
(ii) no employee earns less than £10 000 whereas half of those at Michael's Marketing do.

Each argument is plausible.

Calculation Help desk

James claim	(i)	difference in means: £13 400 – £13 300 = £100
	(ii)	5 out of 20 = 25%, 4 out of 20 = 16%
Sophie's claim	(i)	difference in medians: £11 600 – £9750 = £1850
	(ii)	10 out 20 = 50%

Collecting data

The collection of data can take a variety of forms
• questionnaires
• direct measurement
• counting or observation

Questionnaires: A very common way of collecting data is by questionnaire. You give out a sheet of paper with several questions on it and ask the recipient to write in their answers.

Measurement: Some measuring equipment provides a digital output while others require you to take a reading from a scale. When reading from a scale you need to work out what each sub-division represents.

Example

Find the readings in (i) and (ii) below.

In (i) 2 sub-divisions represent 100.
So 1 sub-division represents $\frac{100}{2}$ =50.
The reading is 300 + 50 = 350.

In (ii) 10 sub-divisions represent 1
So 1 sub-division represents $\frac{1}{10}$ = 0.1
The reading is 31 + (7 × 0.1) = 31.7

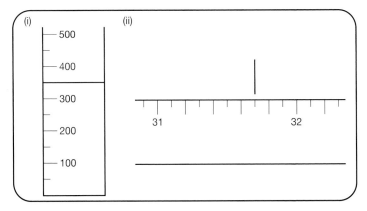

Question 1
Write down
(i) the length of the screw,
(ii) the diameter of the head of the screw.

Question 2
Write down the reading on the speedometer in
(i) miles per hour   (ii) kilometres per hour

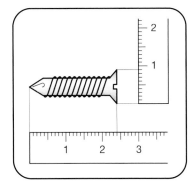
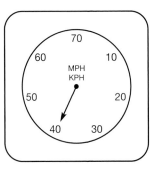

Answers:

1 (i) 2.4 cm.   (ii) 0.9 cm
2 (i) 40 mph   (ii) 65 kmph

Observation: collecting information by watching and recording.

Example:

Amy lives on a busy road that does not have a pedestrian crossing of any sort.

She is trying to persuade the council to build a crossing and is collecting data to support her case. She counts how many people cross the road and how many vehicles use the road during certain periods of time. Amy is collecting data by observation.

Grouping data

Jessica travels to work by train. Unfortunately the train is often late. Jessica decides to collect some data with which to confront the train company. Each day over a period of 6 weeks (30 working days) she records how many minutes late the train is. Here are her results.

```
 0   6 31 15 0      12 38 2 17   1     23  0 14 18 3
22 13   1   7 0       6 11 5 18 25      9  1 13 27 9
```

These results can be summarised in a frequency table as follows.

Minutes late		Number of days		Minutes late		Number of days	
0	11	4	1	20	31	0	1
1	12	3	1	21	32	0	0
2	13	1	2	22	33	1	0
3	14	1	1	23	34	1	0
4	15	0	1	24	35	0	0
5	16	1	0	25	36	1	0
6	17	2	1	26	37	0	0
7	18	1	2	27	38	1	1
8	19	0	0	28	39	0	0
9		2		29		0	
10		0		30		0	

This is not satisfactory. The frequency table has too many classes (one for each number of minutes late). A better approach is to group the data (i.e. group several outcomes together) when producing a frequency table.

This can be done in different ways. One way is to form groups for 0 to 4, 5 to 9, ..., 35 to 39.

The frequency table would look like this.

Minutes late	Tally	Frequency
0 – 4	卌 IIII	9
5 – 9	卌 I	6
10 – 14	卌	5
15 – 19	IIII	4
20 – 24	II	2
25 – 29	II	2
30 – 34	I	1
35 – 39	I	1
Total	30	30

This table is more compact. It is easier to interpret the data.

Other groupings can be used,
e.g. 0 to 9, 10 to 19, 20 to 29 and 30 to 39

Question

Use the classes 0 to 9, 10 to 19, 20 to 29 and 30 to 39 to produce a frequency table.

Answer:

Minutes late	Tally	Frequency
0 –9	卌 卌 卌	15
10 –19	卌 IIII	9
20 – 29	IIII	4
30 – 39	II	2
Total	30	30

Help desk

A quick way to do this is to add two groups together like this
$0 - 4 = 9$, $10 - 14 = 5$
$5 - 9 = 6$ $15 - 19 = 4$
$0 - 9 = 15$ $10 - 19 = 9$

Reading and interpreting information

Tables

Tables are a good way to show numerical information. The numbers are lined up in rows and columns. This makes it easy to find a particular number by referring to the row and column headings.

The table below is displayed in a DIY store. It shows the cost of having your purchases delivered to your home. The delivery cost depends on how much you spend and also on how far you live from the store.

HOME DELIVERY CHARGES			
Amount spent	**Distance from Store to Home**		
	up to 5 miles	**from 5 to 10 miles**	**from 10 to 20 miles**
up to £50	15	20	25
from £50 to £100	10	15	20
from £100 to £200	5	10	15
over £200	free	5	10

Liam lives 3 miles from the store. He spends £189 on a garden shed.

The cost of home delivery is £5.

Question

Work out the cost of home delivery in the following cases.
(i) Diana spends £89 on garden fencing.
     She lives 12 miles from the store.
(ii) Gareth spends £120 on a door.
     He lives 8 miles from the store.
     Answer: (i) £20   (ii) £10

Calculation *Help desk*

1 (i) The amount spent is between £50 and £100.
  The distance is between 10 and 20 miles.
  So the cost is £20.
 (ii) The amount spent is between £100 and £200.
  The distance is between 5 and 10 miles.
  So the cost is £10.

Charts

There are many different types of chart. You need to be able to interpret information presented in bar charts and pie charts.

Pie charts

Pie charts are used to show proportion.

Emma is finding out whether her diet is healthy. She finds this pie chart, which shows the make up of a healthy diet. Emma measures the angle at the centre of the chart for milk and dairy products. It is 60°.

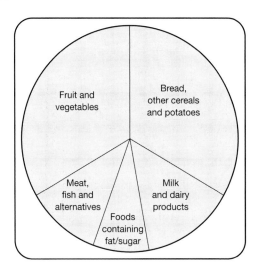

Remember: there are 360° in a circle

To find how much of her diet should be milk and dairy products she needs to work out what 60° is as a fraction of 360°

$$\frac{60}{360} = \frac{1}{6}$$

Emma needs to make sure that $\frac{1}{6}$ of her diet contains milk and dairy products.

Rather than measure the angle Emma may prefer to make an estimate of the proportion of milk and dairy products.

Using fractions she judges that this is about $\frac{1}{5}$. (Don't worry that the estimate differs slightly from the exact calculation – after all it is only an estimate!)

Using percentages she judges that it is about 15% – 20%.

Question

Estimate what proportion of a healthy diet is fruit and vegetables.

Use (i) a fraction   (ii) a percentage

Answer: (i) about $\frac{1}{3}$ (ii) about 30%

Help desk

1 (i) It is about $\frac{1}{3}$. It is certainly more than a $\frac{1}{4}$.

(ii) It is about 30%. It is a little more than 25% but much less than 50%.

Bar chart

Bar charts are used to make comparisons.

Robbie is choosing a hotel for his holiday in Majorca.
He is using this chart to help him make up his mind.
The chart shows the percentage of holidaymakers last year who were happy with (i) food, (ii) accommodation, (iii) hotel staff and (iv) entertainment.

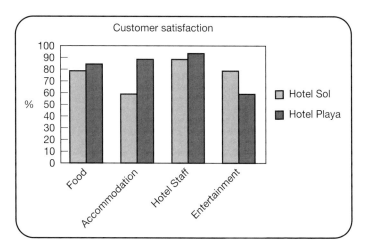

You can read off the percentages for each hotel and compare them.

For Hotel Sol: 80% were happy with food
                 60% were happy with accommodation
                 90% were happy with the hotel staff
                 80% were happy with entertainment.

Question 1

What percentage of holidaymakers at Hotel Playa were happy with (i) food, (ii) accommodation, (iii) hotel staff and (iv) entertainment?

Question 2

Compare the two hotels.

Answers:
1   (i) 85%   (ii) 90%   (ii) 95%   (iv) 60%
2   At Hotel Playa the food and hotel staff were each rated a little better, and the accommodation was much better. At Hotel Sol the entertainment was much better.

(Help) desk

Robbie's decision will depend partly on these factors and how important each is to him. It will also depend on other factors such as cost and location.

Graphs

You need to be able to interpret information presented in line graphs and scatter diagrams.

Line graph

Line graphs are used to show changes over time.

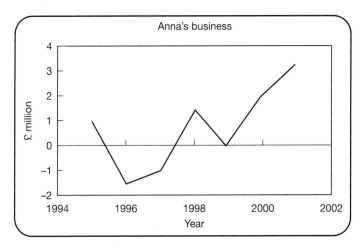

Scatter graph

Scatter graphs are used to show the relationship between two variables.

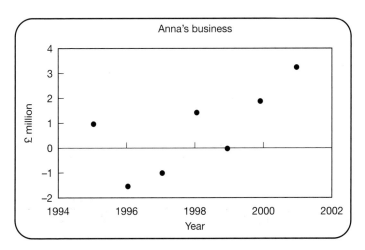

Example

The graphs above show the profits of Anna's Business from 1995 to 2001.

Find the profit (in figures) for (i) 1995 and (ii) 1996.

(i) The profit in 1995 was £1m, i.e. £1 000 000.

(ii) The profit in 1996 was –£1.5m i.e. –£1 500 000.
     This means there was a loss of £1 500 000.

■ NOTE: In business accounts a loss is usually written in brackets.

So a loss of £1.5m would be appear as (£1 500 000).

Question 1

Find the profit or loss for each of the following years.

(i) 1997   (ii) 1999   (iii) 2001

Answers:

1 (i)   The reading is –£1 million. The loss is £1 000 000.

  (ii) The reading is £0. There is no profit or loss (i.e. break-even).

  (iii) The reading is £3.25 million. The profit is £3 250 000.

This graph has **two** lines. The graph shows the percentage of males and females smoking cigarettes over a 30 year period.

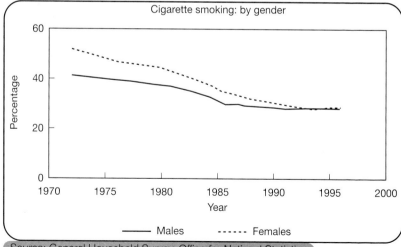

Cigarette smoking: by gender

Source: General Household Survey, Office for National Statistics

Question 2

a) Find the percentage difference between male and female cigarette smoking in
   (i) 1975   [10%]
   (ii) 1995   [1%]

b) Compare the results for the 2 years.

Calculation Help desk

a) From the graph:
   (i) Males 40% , females 50%
       Percentage difference: (50 – 40)% = 10%
   (ii) Males = 29%, females 30%
       Percentage difference: (30 – 29)% = 1%

b) In 1995 females smoked 10% more cigarettes than males. By 1995 the figure dropped to 1%.

Diagrams

The diagrams you are most likely to meet in tests are plans or maps. To interpret them you must be able to use a **scale**.

Example

This map has a scale of 2 cm:1 km.

How far is it from H to W by road?

On the map HL is 5.6 cm and LW is 3.4 cm.

So HW is 5.6 cm + 3.4 cm = 9.0 cm.

The scale is: 2 cm represents 1 km.

This means 1 cm represents 0.5 km (dividing each measurement by 2).

So 9 cm represents 4.5 km (multiplying each measurement by 9).

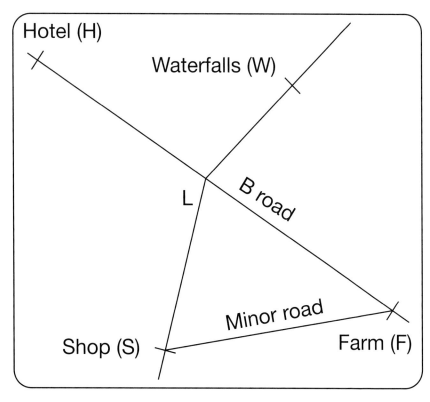

Question

Find the length of the journey from F to S
(i) on the minor road
(ii) on the B road.

Answer: (i) 3.2 km   (ii) 5.45 km

Help **desk**

(i) On the map FS is 6.4 cm.
     2 cm represents 1 km so 1 cm represents 0.5 km.
     Therefore 6.4 cm represents 6.4 × 0.5 km = 3.2 km.
(ii) F to L is 6.2 cm and L to S is 4.7 cm.
     Therefore F to S on the B road is 6.2 cm + 4.7 cm = 10.9 cm.
     1 cm represents 0.5 km.
     Therefore 10.9 cm represents 10.9 × 0.5 km = 5.45 km.

Example

The diagrams below show the front elevation and side elevation of a table.

The scale is 1:20

What are the measurements of the table top?

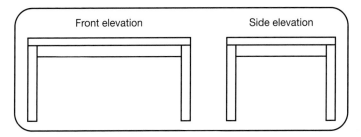

On the plans the length (from front elevation) is 4.5 cm and the width (from side elevation) is 3.0 cm

Multiply each of these by 20 to get the actual measurements.
4.5 cm × 20 = 90 cm and 3.0 cm × 20 = 60 cm

The dimensions are 90 cm by 60 cm

Question

Using the front and side elevations shown above find
(i) the height of the table
(ii) the area of the table top
(iii) the dimensions of a table leg

    Answer:
    (i)   44 cm
    (ii)   5400 cm^2
    (iii) 5 cm × 5 cm

Help desk

(i) Height = 2.2 cm   2.2 × 20 cm = 44 cm
(ii) Area = length × width =   90 × 60 cm^2 = 5400 cm^2
(iii) Table leg is 0.25 cm   0.25 × 20 cm = 5 cm

7

Practice for the Application of Number Test at Level 2

Introduction

The test has 40 multiple-choice questions.

In each question there is a context.

The text of the question describes a situation, or gives you information about a person or organisation.

You are then given four possible answers, A, B, C and D.

You have to choose the correct option, then mark the letter of your choice on an answer grid.

DON'T JUST GUESS – WORK IT OUT

Don't expect to be able to work it out in your head without writing anything down.

You should be given some rough paper on which to do the calculations. If not, ask for some. If there isn't any, then use the margins in the question booklet.

How to tackle the paper

Work through the paper, attempting all the questions, in order. If you do not understand one of the questions, mark it on the paper (e.g. with a circle or star), and go on to the next question. When you get to the end of the paper, go back to the beginning. Read through again, looking for any you missed on the first run through.

Have another go.

If you can't decide which calculation to do,

Try to think of the scene being described – what sort of number are you looking for?

- If you expect the answer to be a large number, you need to add or multiply some of the numbers you are given
- If you expect the answer to be a small number, you need to subtract or divide some of the numbers you are given

Try to judge the size of the answer – should it be units, tens, hundreds, etc.

If all else fails it is better to guess than to leave a blank. You have a 25% chance of guessing correctly. That's better than the lottery!

 If you still have time, check through all your answers again.

How to tackle each question

Read through all of the information, so that you know what the question is about.

Read the sentence that contains the actual question carefully. Pick out the key words. Underline them if it helps to focus your mind.

Go back to the information and find the numbers you need for your calculation.

Write out the calculation.

Check your result with the options you are given: A, B, C and D, and decide which letter gives the correct answer.

Mark this letter on the answer grid.

If you can't find your result among the options A, B, C or D

You made a mistake! Check

- Did you do the correct calculation?
  (e.g. if you added two numbers, should you have multiplied them?)

● Did you carry out the calculation correctly?
   (e.g. did you remember your multiplication tables correctly?)
Go through the question again, or, if you think you are short
of time, go on with the rest of the test, but mark this question
to check again at the end.

Remember, the *wrong* options are the numbers you get if
you make a mistake in your calculations.
Nasty!

Use all the time you have.
Check each question.
Check the whole paper.

In the next section you will find a sample test paper. There are
21 multiple choice questions. These are all based on
questions taken from past papers at Level 2. You should be
able to complete all 20 in half an hour.

Do the paper in test conditions.

Sit somewhere quiet.
Give yourself half an hour.
Make sure you have some rough paper to do the calculations
on.
No calculator!
No asking for help!

Good luck.

SAMPLE TEST PAPER

1   A nurse is taking a patient's temperature.

The thermometer measures in degrees centigrade (°C).

What is the reading?

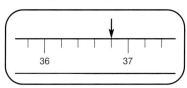

A 36.4 °C
B 36.8 °C
C 36.9 °C
D 37.1 °C

2 Matthew is helping his daughter learn about fractions, decimals and percentages.

The fractions, decimals and percentages in each row of this table are meant to be equal.

Row	Fraction	Decimal	Percentage
1	$\frac{1}{10}$	0.1	10%
2	$\frac{1}{4}$	0.4	40%
3	$\frac{1}{2}$	0.5	50%
4	$\frac{9}{10}$	0.9	90%

There is an error in the table.

In which row does the error occur?

A Row 1
B Row 2
C Row 3
D Row 4

Questions 3 and 4 are based on the following.

Vicky's television set cost £364.55

It weighs 30 kg

The screen size is 52 cm

3 What is the cost of Vicky's television set, to the nearest ten pounds?
   A £360.00
   B £364.00
   C £365.00
   D £370.00

4 Given 1 kg is about 2.2 lbs and 1 cm is about 0.4 inches. Which of the following statements is true?
   A The set weighs about 15 lbs.
   B The set weighs about 66 lbs.
   C The screen size is about 21 inches.
   D The screen size is about 15 inches.

5 One day Ashley recorded the temperature every two hours. This is the graph he drew.

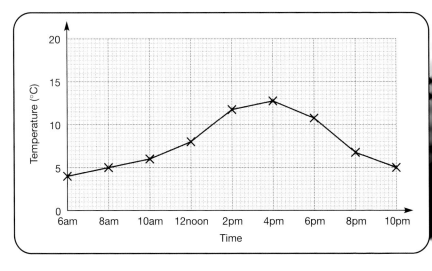

Which of the following statements is true?
   A The time period covered by the graph is 4 hours.
   B The highest temperature was recorded at 4pm.
   C The lowest temperature was recorded was at 10pm.
   D The temperature at 10am was 4 °C.

6 Harriet earns £4.00 an hour. She works 8 hours a day. She works 5 days each week.
   Which of the following statements is true?

A Harriet earns £1 for every 25 minutes that she works.
B Harriet earns £28 on each day that she works.
C Harriet works 35 hours each week.
D Harriet earns £160 each week.

7 Here is a plan of an L-shaped office.

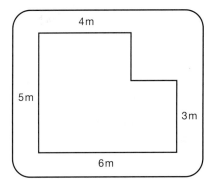

What is the area of the office?
A 18 m²
B 22 m²
C 26 m²
D 38 m²

Questions 8 and 9 are based on the following:

Stuart asked his friends how they travelled to college.
The results are shown below.

Method of travel	Number
Car	10
Train	1
Motorcycle	6
Bus	17
Bicycle	4
Walk	2
Total	40

8 Which of the following statements is true?
   A More friends travelled by train than motorcycle.
   B Twice as many friends walked as travelled by bicycle.
   C More than half his friends travelled by bus.
   D A quarter of his friends travelled by car.

9 Stuart draws this pie chart to represent the data but has
   not labelled it.

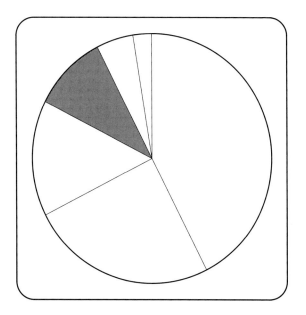

   Which method of travel is represented by the shaded sector?
   A Car
   B Motorcycle
   C Bicycle
   D Walk

10 The map shows an island.

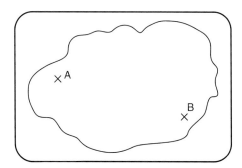

The scale of the map is 1:50 000

Jessica wants to work out the distance between landmarks A and B

In real life, what is the approximate distance between A and B?

A 1250 m

B 2000 m

C 125 000 m

D 200 000 m

11 Gabriella buys a computer printer priced at £200 + VAT.

VAT is added at $17\frac{1}{2}\%$.

How much does Gabriella pay when VAT is included?

A £165.00

B £182.50

C £217.50

D £235.00

12 A circular metal disc has a diameter of 20 mm.

Paul wants to find the area.

The area of a circle is πr^2 where r is the radius.

Taking π as 3 what is the area of the disc?

A 300 mm²

B 600 mm²

C 900 mm²

D 1200 mm²

13 Josh is designing a rectangular playground with an area of 2400 m².

The length, L, is given by the formula $L = \dfrac{2400}{W}$.

Which value of W gives the largest value of L?

A $W = 10$

B $W = 20$

C $W = 30$

D $W = 40$

14 In an examination Sunil scores 60 marks out of 75.

What is Sunil's mark as a percentage?

A   45%

B   60%

C   75%

D   80%

15 This table shows the cost of posting a letter.

Weight (g)	First class	Second class
up to 60	27 p	19 p
60 – 100	41 p	33 p
100 – 150	57 p	44 p

Jo posts two letters each weighing 50 g first class and one letter weighing 90 g second class.

What is the total cost of postage?

A   60 p

B   73 p

C   87 p

D   93 p

Questions 16 and 17 are based on the following.

A group of fifty people were asked how many lottery tickets they had bought last week.

The results are shown in the table.

Number of tickets	0	1	2	3	4	5	6	7	8
Number of people	18	8	10	5	3	2	3	0	1

16 What percentage of the group bought no lottery tickets?

A   7

B   9

C   18

D   36

17 Which of the following statements about the tickets is true?

A   The mean is 4.

B   The median is 1.

C   The mode is 4.

D   The range is 9.

Questions 18 and 19 are based on the following.

A rectangular plot of Emma's garden is to be concreted to create a parking space for her car. The concrete is to be 10cm deep.

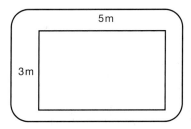

18 What is the perimeter of the plot?
    A 8m
    B 13m
    C 15m
    D 16m

19 What volume of concrete is required?
    A 0.015m³
    B 0.15m³
    C 1.5m³
    D 15m³

20 A DIY shop sells white gloss paint in 4 different sized tins. They are priced as shown.

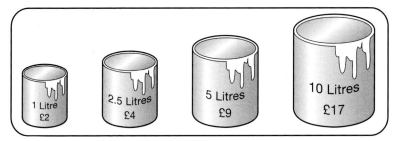

Which tin gives the lowest cost per litre?
    A The 1 litre tin.
    B The 2.5 litre tin.
    C The 5 litre tin.
    D The 10 litre tin.

21 This bar chart shows the grade of the 100 employees of PJD Insurers.

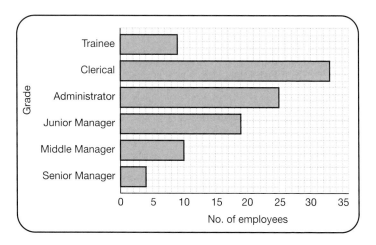

Which of the following statements is true?

A   There are 21 Junior Managers.

B   There are more than 30 Administrators.

C   The are fewer than 10 Trainees.

D   There are 4 employees on Middle Manager grade.

⑧ Apply Your Skills at Level 2

Understanding Part B

Part B of the specifications is about the **application** of number skills in the context of a specific activity, which may be part of your main programme of study.

You will need to build a 'portfolio' of evidence showing how you have met all the requirements of the Level 2 Application of Number specifications.

- All components of Part B of the specifications must be covered
- All the assessment criteria must be fully met

At this level you must be able to carry through a *substantial activity* that requires you to:

Interpret information:

- Decide how to find the information that you will need
- Obtain the relevant information
- Decide what calculations you will need to use to get the results you need

Use this information when carrying out calculations:

- Carry out calculations involving two or more steps and numbers of any size
- Clearly show your methods and the levels of accuracy you are working to
- Check your methods to identify and correct any errors
- Make sure that your results make sense

Interpret your results and present your findings:

- Select effective ways of presenting your findings
- Present your findings clearly
- Describe the methods you used
- Explain how the results of your calculations meet the purpose of the activity

You are expected to carry out at least one *substantial* activity that can be broken down into tasks for each component of the unit. You must make sure that:

- All calculations are clearly set in context
- All calculations are clearly set out, with evidence of checking
- You show that you are clear about the purpose of your task

Definition Help desk

- **Substantial activity** – an activity that includes a number of related tasks where the results of one task are required to carry out subsequent tasks.

  For example, an activity will involve obtaining and interpreting straightforward information, using this information when carrying out calculations, and interpreting the results of the calculations in the context of presenting findings

- Straightforward subject – a subject that you may often meet in your work or studies

Portfolio building

A portfolio is usually a file or folder that you use to present the evidence that shows how you have met the requirements for Part B of the Application of Number specifications. Portfolio building is very important. You will need to plan and organise your work from the start.

Portfolio Help desk

Your portfolio should include:

Portfolio documentation that must:

- state which aspects of the unit you are claiming
- state exactly what work you have produced
- describe how this work covers the relevant parts of the unit
- state where the evidence can be found by the use of a clear referencing system
- be signed by an appropriate member of staff

Assignment briefs or tasks

Evidence of achievement showing:

- You have planned and carried through at least one substantial activity that included tasks for N2.1, N2.2 and N2.3
- Shorter additional activities to cover any outstanding evidence (e.g. types of calculations, source of information or forms of graphical presentation)

Start to build your portfolio as soon as possible. You can always remove work later if you find that you have produced something better.

Why do I need portfolio documentation?

- To help you to organise your work
- To make it easy for a moderator to see that you have sufficient evidence
- To make it easy for a moderator to find the evidence
- To show a moderator where to find pieces of evidence that are not kept in your portfolio

What can I use as evidence?

- Evidence can be taken from a range of contexts (e.g. your programme of study, work experience, community activities and voluntary work.)
- Evidence can be hand-produced or electronically produced and may include:
  - Written material, including calculations
  - Draft material showing the preliminary calculations, etc.
  - Visual material such as graphs, charts and diagrams
  - 3-D scale models
  - Records of information that you have obtained, e.g. surveys
  - Copies of source materials, e.g. data from the internet

Quality is more important than quantity.

Definition Help desk

- **Moderation**
  The reassessment of a sample of portfolios to check that work is being assessed accurately and consistently to the National Standards.

- **Internal moderator**
  A key skills specialist appointed in your school/college/training agency. The internal moderator checks that your evidence has been assessed accurately and consistently to the National Standards.

- **External moderator**
  A person appointed by the Awarding Body. The external moderator checks that your evidence meets the agreed National Standards.

Providing the evidence

Evidence Help desk

You must:

- Decide how to find the information that you need
- Find the relevant numerical information needed to carry out the task
- Interpret the information correctly
- Make appropriate calculations accurately
- Present findings in written and graphical forms

Example

Four friends decide to go on holiday. They cannot decide where to go. They have narrowed the choice down to Spain, Ibiza, Greece or one of the Greek islands. They want you to research two destinations for them and present your findings with your recommendations.

They have agreed that the holiday must cost less than £300 per person.

Two want to spend their time on the beach or by the pool enjoying the sunshine and working on their tan. The other two are more energetic and want to be able to take part in outdoor activities. All four want plenty of nightlife.

All the countries are within the Euro zone. You will need to provide them with information about the Euro. They will need to know how to work out how many Euros they will get for £1, £10, £20, etc. Use appropriate statistical measures to compare the exchange rates between the pound and the Euro over a 21 day period last year with a similar period in the current year.

The weather is important for a good holiday. Use appropriate graphs and charts to present information about temperatures, rainfall, etc.

Prepare a report of your findings and recommendations. Include all your calculations. Use graphs, charts and diagrams where appropriate.

How do I begin?

INTERPRET INFORMATION

- Investigate destinations:
  - ○ Travel agents
  - ○ Internet
  - ○ Specialist companies
- Investigate the weather
  - ○ Temperatures
  - ○ Sunshine
  - ○ Rainfall
- Type of holiday:
  - ○ Outdoor activities
  - ○ Beach
  - ○ Nightlife
- Type of accommodation:
  - ○ Hotel
  - ○ Self-catering

Evidence should include:

- A clear explanation of the purpose of the activity
- How you used 2 different sources to find information that was relevant
- Information you obtained from **2 different types** of sources
- How you decided what methods you would use to get the required results
- Notes of how you made your choices
- Copies of source material
- Records of the information you obtained

What do I do next?

CARRY OUT CALCULATIONS

- Costing calculations
  - ○ Cost of travel
  - ○ Cost of accommodation
  - ○ Supplements/reductions

- Currency exchange rates
  - Calculate mean, median and range for exchange rates over at least 20 days
  - Compare mean, median and range with given data
- Weather information from graphs
  - Temperatures
  - Sunshine
  - Rainfall

Evidence should include:

- Detailed calculations involving two or more steps

- At least one example from each of the 4 categories

- Comparison of two sets of data with a minimum of 20 items

- Records of your calculations that clearly show the methods and levels of accuracy you used

- Notes of how you checked methods, corrected any errors and made sure your results made sense

---

How do I present my findings?

INTERPRET RESULTS OF CALCULATIONS AND PRESENT FINDINGS

- Decide how you will present your findings
- Remember to include detailed costing calculations
- Use graphs, charts and diagrams to present information
- Make your recommendations

Examples of graphs, charts and diagrams

- Graphs
  - Exchange rates
  - Currency conversion
- Charts
  - Sunshine
  - Rainfall/snowfall
  - Temperature
- Diagrams
  - Scale drawing of accommodation
  - Map of area
  - Flowchart for travel schedule

Evidence should include:

- Descriptions of the methods used to obtain your results
- Explanations of the results of calculations in terms of how they met the purpose of the activity
- Selection of appropriate forms of presentation to match the information being presented
- Effective presentation of findings. Use of at least one graph, one chart and one diagram

The presentation must be of 'publishable quality'

Some source data

Exchange Rate Information
British Pound to Euro
August 1st – 21st 2000

Mean	1.6513
Median	1.6505
Range	0.0531
Minimum	1.6184
Maximum	1.6715

Exchange Rates – British Pound to Euro

August 1st to 21st 2001

01.08.2001	1.6293	08.08.2001	1.6149	15.08.2001	1.5838
02.08.2001	1.629	09.08.2001	1.6117	16.08.2001	1.5792
03.08.2001	1.6217	10.08.2001	1.5999	17.08.2001	1.584
04.08.2001	1.6172	11.08.2001	1.5953	18.08.2001	1.5762
05.08.2001	1.6166	12.08.2001	1.5931	19.08.2001	1.5746
06.08.2001	1.6166	13.08.2001	1.5931	20.08.2001	1.58
07.08.2001	1.6106	14.08.2001	1.5849	21.08.2001	1.5813

Source: www.oanda.com

Resort Information
Source: http://members.tripod.co.uk/HolidayWeather

IBIZA

	Jan	Feb	Mar	Apr	May	Jun	Jul	Aug	Sep	Oct	Nov	Dec
Precipitation Days	6	6	6	5	3	3	1	1	4	8	10	8
Sunshine hours	5	6	7	8	10	11	11	11	8	6	5	5
Warm Clothing	✓	✓	✓									✓
Outdoor Activities		✓	✓	✓	✓	✓	✓	✓	✓	✓		
Sunbathing / Swimming					✓	✓	✓	✓	✓			
Recommended					✓	✓	✓	✓	✓			

BENIDORM

	Jan	Feb	Mar	Apr	May	Jun	Jul	Aug	Sep	Oct	Nov	Dec
Rainy Days	3	2	4	5	3	2	1	1	6	5	6	4
Warm Clothing	✓	✓	✓	✓							✓	✓
Outdoor Activities	✓	✓	✓	✓	✓	✓	✓	✓	✓	✓	✓	✓
Sunbathing / Swimming						✓	✓	✓	✓	✓		
Recommended					✓	✓	✓	✓	✓	✓		

GREECE INCLUDING THE ISLANDS

	Jan	Feb	Mar	Apr	May	Jun	Jul	Aug	Sep	Oct	Nov	Dec
Precipitation Days	12	9	8	4	3	1	0	0	1	4	6	12
Daily Sunshine Hours	3	3	5	6	8	11	13	12	10	6	3	3
Warm Clothing	✓	✓										
Outdoor Activities		✓	✓	✓	✓	✓	✓	✓	✓	✓	✓	✓
Sunbathing / Swimming					✓	✓	✓	✓	✓	✓		
Recommended						✓	✓	✓	✓	✓		

Weather graphs

Source: http://members.tripod.co.uk/HolidayWeather

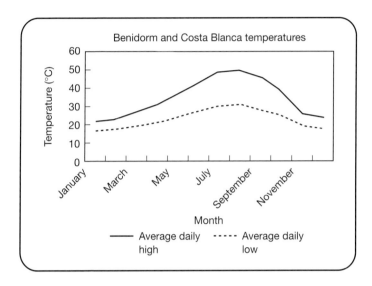

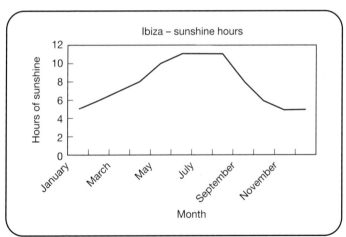

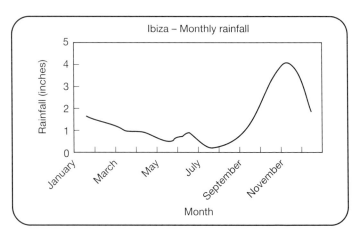

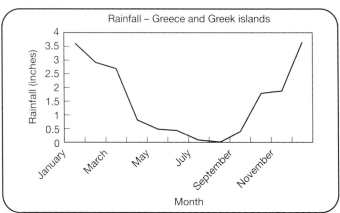

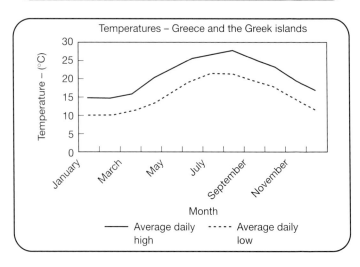

Introduction

This part of the book will help you to develop and practice the Application of Number skills as set out in Part A of the unit. It will also help you to get ready for the test at Level 3.

- The test is one-and-a-half hours long
- It has both short answer and extended questions
- The questions test both Part A and Part B

At Level 3 you must be able to plan and carry through a substantial and complex activity that requires you to:

- Plan your approach to obtaining and using information, choose appropriate methods for obtaining the results needed, and justify your choice
- Carry out multi-stage calculations, including use of a large data set (over 50 items) and re-arrangement of formulae
- Justify your choice of presentation methods and explain the results of your calculations

You are expected to demonstrate your skills in the context of at least one substantial and complex activity that can be broken down into a series of interrelated tasks covering all three components of the unit.

- Multi-stage calculations should be carried out
- All calculations must be clearly set in context
- You must show that you are clear about the purpose of each activity

• You must show that you have considered the nature and sequence of tasks when planning how to obtain and use information to meet your purpose

Understanding Part A

Read through the list of 'bullet points' in Part A of the specification. There may be things you think you can do and some things you are not sure about. If you have not already done so, turn to the **self-assessment check-list** on page 15 and complete it. To pass the test you need to be sure that you can do all of Part A. The checklist will help you to identify the skills that you have and those that you need to work on.

Numbers and Calculating

Using a calculator at Level 3

At Level 3 you will need a **scientific calculator**, not just a simple one.

This is because of the more complicated calculations you will have to do – where the extra function keys, like x^y or ( ) or $\frac{1}{x}$ or *tan* will make calculations much easier to carry out in your portfolio. You will have to know how to use these function keys for your external test.

Your calculator will also help you understand or check new calculations you do in your head or on paper.

It may not have a % key, though, so if you like to use a % key, keep a simple calculator to hand also.

A scientific calculator looks like this. . . . .
compared with this.

Getting used to your scientific calculator

A good way to get used to how your calculator works is to try simple calculations on it – ones that you know the answers to. Then you can see what the calculator is doing when you press particular keys. **There are slight differences between the ways calculators work, so make sure you have the instructions to hand.** The more you experiment with your calculator, the more you will learn how it works.

Calculator Help desk

Remember

- Don't forget to clear your calculator before you start a new calculation
- Switching off and on clears everything
- Pressing AC clears your last calculation but doesn't clear the memory
- Pressing C or DEL just clears the last number you entered
- Press = to finish off a calculation

Use your calculator to work out these questions

(i) $12 - 4 \times 2$ [4]

(ii) $12 \times 2 - 4 \div 2$ [22]

(iii) $12 \div 3 + 4 \times 4 - 1$ [19]

Using brackets

Some calculations will involve brackets.

Brackets make clear which parts of a calculation to do first.

Operations Help desk

Rule 1 Work from left to right

Rule 2 Work out × and ÷ before working out + and −

Rule 3 Do the calculations inside brackets first

Calculator Help desk

Numbers	Work out 4 + 3 − 5
	Press the keys 4 + 3 − 5 =
	The calculator display is 2
Fractions	Work out $2\frac{1}{4} + 3\frac{1}{8}$
	Press the keys 2 $a^{b/c}$ 1 $a^{b/c}$ 4 + 3 $a^{b/c}$ 1 $a^{b/c}$ 2=
	The calculator display is $5\frac{3}{8}$
Percentages	Work out 17.5% of £225
the percentage key	Press the keys 230 × 17.5 shift % =
is often above the = sign	The calculator display is 40.25
Brackets	Work out 3(4 + 5 − 3)
	Press the keys 3 × (4 + 5 − 3) =
	The calculator display is 18

Try these calculations on your calculator.

Planning calculations

Look at this shopping list

Description	Quantity	Unit price before VAT
Time switch	3	£11.89
23w long life bulb	4	£8.40
13A plug	2	£1.79

Think of different ways to work out the total cost, including the VAT at 17.5%. Don't work it out yet! There are lots of alternative ways.

Note down three of the ways now!

Three possible methods:

Method 1
- Work out the unit price of each item including the VAT
  £11.89 + (£11.89 × 17.5 ÷ 100)
- Work out the total cost of each item
- Add up the three totals.

Method 2

- Work out the cost of each line before VAT (such as $3 \times £11.89$)
- Add on the VAT for that line
- Add the three line totals

Method 3

- Leave the VAT calculation until the end.
- Work out the total cost of all the items without VAT
- Add on the VAT

HINT: Make a rough estimate to check your calculations

Here is an estimate following method 3.

Estimation Help desk

Step 1.   Round all the money amounts to the nearest £.

£11.89 rounds up to £12

£8.40 rounds down to £8

£1.79 rounds up to £2

Step 2.   Calculate the total excluding VAT

$3 \times £12 = £36$

$4 \times £8 \ = £32$

$2 \times £2 \ = £4$

Total $= £72$

Step 3.   Estimate the VAT

17.5% is more than 10% (£7.20) but less than 20% (£14.40) – say roughly £10

Step 4.   Add the VAT £72 + £10 = £82

The estimated cost is approximately £82

Using your calculator for Method 3

Calculation Help desk

To calculate cost without VAT

Press the keys	$3 \times 11.89 + 4 \times 8.4 + 2 \times 1.79 =$	$3 \times 11.89 + 4 \times 8.4 + 2 \times 1.79 =$
Display shows	72.85	72.85
Now press	M+	Min

To calculate the VAT

Press the keys	RCL M+ = × 17.5 shift %	MR × 17.5 shift % = M+
Display	12.74875	12.74875

To add the VAT

Enter	+	+
Display	85.59875	85.59875

Rounding to 2 decimal places, the total cost, including VAT, is £85.60

Use your calculator to work out the cost using methods 1 and 2 – and your own way also, if it is different from all of these. All three methods give the same total – £85.59875 which rounds to £85.60

Remember: clear your calculator before you start a calculation

Powers and Roots

Squares and cubes
You will often come across numbers raised to the power 2, such as 4^2

    Say "4 to the power of 2"

    4^2 means 4×4

Usually you say these powers to the 2 as 'squared'

    '5 squared' means 5^2,

    52 means 5×5

You will also come across numbers raised to the power 3, such as 5^3

    5^3 means $5 \times 5 \times 5$

    Say "5 to the power of 3"

Usually you say these powers to the 3 as 'cubed'

    '10 cubed' means 10^3.

    10^3 means $10 \times 10 \times 10$

Question

What is the value of 7 squared + 4 cubed?   [49 + 64 = 113]

Square roots and cube roots

The opposite (or 'inverse') of squaring is finding the square root
$7^2 = 49$ so the square root of 49 is 7
The opposite (or 'inverse') of cubing is finding the cube root
$4^3 = 64$ so the cube root of 64 is 4

The sign for square root is $\sqrt{}$

$\sqrt{64} = 8$

The sign for cube root is $\sqrt[3]{}$

$\sqrt[3]{125} = 5$

Question 1

What is
(i) the square root of 81?   [9]
(ii) the cube root of 125?   [5]
(iii) the square root of 64?   [8]
(iv) 6 to the power of 4?   [$6 \times 6 \times 6 \times 6 = 1296$]

Question 2

What are the values of
(i) $\sqrt{16}$ [4]
(ii) $\sqrt[3]{1000000}$ [100]
(iii) $\sqrt{144}$ [12]

Calculator Help desk

Squares	Work out 13^2		
	Press the keys	$13\ x^2\ =$	
	Display shows	169	
Cubes	Work out 13^3		
	Press the keys	$13\ x^3 =$	
	Display shows	2197	
Square Root	Work out $\sqrt{169}$		
	Press the keys	$\sqrt{\ }169 =$	$169\ \sqrt{\ } =$
	Display shows	13	
Cube Root	Work out $\sqrt[3]{2197}$		
	Press the keys	$\sqrt[3]{x}\ 2197 =$	$2197\ \sqrt[3]{x} =$
	Display shows	13	
Powers	Work out 5^6		
	Press the keys	$5\ x^y\ 6 =$	
	Display shows	15625	
Roots	Work out $\sqrt[4]{14641}$		
	Press the keys	$4\ ^x\sqrt{y}$ shift $14641 =$	14641 shift $^x\sqrt{y}\ 4 =$
	Display shows	11	
Negative powers	Work out 5^{-2}		
	Press the keys	$5\ x^y - 2 =$	$5\ x^y\ 2 ^+/_- =.$
	Display shows	0.04	
	$(1 \div 25 = 0.04)$		

As you multiply again and again by the same number, the power goes up by one each time

$$5^3 \times 5\ (= 125 \times 5) = 5^4\ (=625)$$
$$10^4 \times 10\ (= 1000 \times 10) = 10^5\ (=100\,000)$$

In reverse, as you divide again and again by the same number, the power goes down by one each time

$$5^4 \div 5\ (= 625 \div 5) = 5^3\ (= 125)$$
$$5^1 \div 5\ (= 5 \div 5) = 5^0\ (= 1)$$

Help desk

Press the keys	$5\ x^y\ 0 =$
Check that the display shows	1

Negative powers

Continuing to divide by 5 the power goes down by one becoming negative

$$5^0 \div 5 = 5^{-1} = 1 \div 5 \text{ or } \frac{1}{5}$$

$$5^{-1} \div 5 = 5^{-2} = \frac{1}{5} \div 5 = \frac{1}{25}$$

Check this on your calculator, you may need to use the +/_ button to change the 2 into a –2

Operations Help desk

Rule 1	Work from left to right
Rule 2	Work out $\times$ and $\div$ before working out + and –
Rule 3	*Do the calculations inside brackets first*
Rule 4	**Work out powers before multiplying or dividing**

Question

Use the rules to work out $(4 - 2.9)\ 5^6$

Calculation Help desk

Rule 3	Do the calculations inside brackets first	$4 - 2.9 = 1.1$
Rule 4	Work out powers before multiplying and dividing	$5^6 = 15625$
Rule 2	Work out $\times$ and $\div$ before working out + and –	$1.1 \times 15625 = 171875$
		$(4 - 2.9)\ 5^6 = 171875$

Very large numbers – Millions, Billions, Kilo, Mega and Giga

You may come across any of these when dealing with very large numbers. It's worth being clear what they all mean.

K – short for 'kilo' – is often used to mean 'thousand'		
	A thousand is ten hundreds	1000
M – short for 'mega' – is often used to mean million		
	A million is a thousand thousands	1000 000
G – short for 'giga' – is often used to mean billion		
	A billion is a thousand millions	1000 000 000

Examples

K A salary of £25K means £25 000

M A power station generating 8 Megawatts means it is generating 8 000 000 watts of electricity.

At the end of the year 2000 over 30 000 Ml (megalitres) of water were flowing down the River Severn each day.

G The transmission frequency of GSM mobile phones is at 0.9 and 1.8 gigahertz

£2350 million – the sort of profit that many multinational companies make in a year is £2350 000 000

30000 Ml is 30 billion litres

£2350 million – is £2.35 billion

Notice that K, M and G are all capital letters.

Using powers of 10 to express very large numbers

You may find $\times 10^3$ used to mean thousands
$£25 \times 10^3$ is £25 000

You may find $\times 10^6$ used to mean millions
8×10^6 watts is 8 000 000 watts

You may find $\times 10^9$ used to mean billions
0.9×10^9 hertz is 900 000 000 hertz
1.8×10^9 hertz is 1 800 000 000 hertz

Very small numbers –
Hundredths, thousandths, millionths, centi-, milli- and micro-

You may come across any of these when dealing with very small numbers. It's worth reminding you what they all mean.

c	– short for 'centi' – is often used to mean 'hundredth' A hundredth is 0.01
m	– short for 'milli' – is often used to mean 'thousandth' A thousandth is 0.001
μ– mc	– short for 'micro' – is often used to mean millionth A millionth is a thousandth of a thousandth It is written 0.000 001
(μ is 'mu' the Greek letter for 'm'.)	

Examples

c   centimetre (cm) – hundredth of a metre
centilitre (cl) hundredth of a litre

m   millimetre (mm) – thousandth of a metre, milligram(mg)
a thousandth of a gram – for instance a drug dose of
5 mg means 0.005 g

μ 5 μs(0.000 005 seconds) is the time it takes light to travel
a mile

mc 5 mcg (0.000 005 grams) is the amount of vitamin D an
adult needs each day (you can get this from oily fish,
eggs or margarine)

Notice that c, m, μ and mc are small letters

Using negative powers of 10 to express very small numbers

You may find '$\times 10^{-2}$' used to mean 'hundredths'
    as $10^{-2} = 0.01$ or 1 hundredth
You may find '$\times 10^{-3}$' used to mean 'thousandths'
    as $10^{-3} = 0.001$ or 1 thousandth
    24×10^{-3} seconds is 24 milliseconds
You may find '$\times 10^{-6}$' used to mean 'millionths'
    as $10^{-6} = 0.000 001$ or 1 millionth
    8×10^{-6} grams is 8 micrograms

Calculator Help desk

Using the **EXP key** Work out 4×10^5
    Press the keys   4 EXP 5 =
    Display shows 400000

    Work out 4×10^{-5}
    Press the keys 4 EXP – 5 =   4 EXP 5 $^+/_-$
    Display shows   4 $^{-05}$

Using powers and roots

Saving money

If you save money in a bank or building society, the interest
(that is the extra they add on) is paid yearly, monthly or every
three months (quarterly). Because the total amount increases
each time, the interest gets more and more as well, even

though the % rate may stay the same – this is because you are receiving interest on a larger and larger total.

Calculation Help desk

You invest £200 at an interest rate of 6% 'per annum'

after 1 year, you will have £200 + 6% of 200 = 200 + 12 = £212

after 2 years, you will have £212 + 6 % of £212 = £224.72

See how the interest after the first year was £12, while after the second year it was £12.72

Check these figures now on your calculator.

You can save yourself time by using this result – still using the 6% interest rate.

Amount + 6% of the amount = 1.06% of the amount

so to find the total after 4 years, all you need to do is work out

$200 \times 1.06 \times 1.06 \times 1.06 \times 1.06$

Do this yourself now, and make sure you get £252.4954. . . = £252.50 to the nearest penny

Try these yourself.

1 How much will you have after 4 years?
   [after 3 years you will have £238.20, after 4 years you will have £252.50]

2 After how many years will you have over £300?
   [after 5 years you will have £267.65, after 6 years 283.70 and after 7 years £300.72]
   [your results may have been very slightly different, if you used amounts rounded to the nearest penny at the ends of each year in the following year's calculation.]

The starting amount is also called the "**principal**".

Formula Help desk

Total after saving at r% a year for n years = (original amount) $\times (1.0r)^n$

Example: Original amount =£2500, rate = 5%, period = 4 years

Total = £2500 x 1.05^4 = £3038.77 to 2d.p.

You can use the same method if the interest is added monthly or quarterly (every 3 months). But now the n stands for the number of months, or quarters, rather than the number of years.

Borrowing

Borrowing works the same way – except that unless you repay in one 'go', each month the amount you owe goes up (because of adding on the month's interest) before you subtract the monthly repayment. Many people face real problems if they have to repay over a long period of time, owing to these increases.

The effects of borrowing can be worrying in other ways. Interest rates are much higher than for saving – for instance many credit card companies charge over 20% a year on amounts owing. And if you miss a repayment of the loan, the amount you owe increases.

Another reason is that people often do not work out just how high the yearly interest rate is. All in all, it is very easy for debts to get out of control.

Saving and inflation

Every year, on average, prices tend to go up a little. This is called inflation. On average, what cost £1.00 last year may now cost £1.03, for instance – an inflation rate of 3%. Inflation is worked out as an average of the changes of many different prices – for housing, food, clothes, travel and many other day-to-day expenses. Of course particular items go up (or down) in price by much more.

This means that if you save at a lower rate than the rate of inflation, the value of your savings (what they will buy) is actually going down.

Shrinking amounts and populations

You may also come across situations where amounts are decreasing by the same percentage, year by year.

Example

The owl population in an area is decreasing year by year. In 1995 a survey showed there were about 150 owls in the area. By 2000 the number had dropped to 90. Estimate the annual percentage decrease (to the nearest 1%) assuming the owl population goes down by the same percentage each year.

Calculation Help desk

You can do this step by step:

- choosing a possible percentage

- seeing what it gives

- then improving your choice.

Example:

- start with a 5% decrease. This corresponds to multiplying by $1 - 0.05$ or 0.95

- After 5 years there would be 150×0.95^5 owls $= 116$ owls. So the decrease is more than 5%

- Try 7% (corresponding to 0.93), which gives $150 \times 0.93^5 = 104$ owls

- Try 9%, which gives 94 owls

- 10% gives 89 owls – so 10% a year is the closest

Carry out multi-stage calculations

Multi-stage calculations are when you have to do one calculation, then use the answer to do another, and so on. For instance, to estimate how best to carpet a room for £200, you will probably have to:

- Work out the area of the floor you want carpeted
- Choose some carpet designs you like
- Multiply this by the cost per square metre (including VAT) of different carpets you would accept, allowing for patterns in the carpet
- Take into account that carpet comes in different widths on the roll
- Add on an amount for an underlay
- Allow for any carpet fitting charges
- Make sure the total cost is under £200 – if not choose a different carpet and repeat

You will find you have to do multi-stage calculations later in this book, for your portfolio and in the external test.

Show your methods clearly; work to appropriate levels of accuracy

When you are doing calculations – particularly multi-stage ones – it's important to be able to check what you have done for mistakes. Later you may well have to convince other people that your calculations are correct – this means explaining them so other people can understand.

All this means that it's important to show your calculations clearly and fully.

It's all right to do 'back of the envelope' notes to start with – but you need to write down calculations more fully in your portfolio – and for the external tests.

Get in the habit of writing down what your figures mean – instead of writing just the figures. This means using words as well as numbers!

> Remember: Always tell the story of your calculation

Example:

In the carpet problem we decide that we need 12m^2 of carpet plus underlay. We choose a flower design that costs £ 8.99 per m^2 and underlay costs £4.50 per m^2.

Telling the story:

    We shall need 12 square metres of carpet.

    The flower design costs £8.99 per square metre.

    Underlay costs £4.50 per square metre.

One square metre of carpet + 1 sq metre of underlay = £8.99 + £4.50 = £13.49

The total cost of carpet + underlay is 12(8.99 + 4.50) = £161.88

This will be easier to understand when you go back to your calculation than:

    12

    £8.99 and £4.50 = 161.88

Measuring, quantities and space

Percentages and Proportional change

There's nothing special about the idea of proportional change
– it's a feature of daily activities for nearly everyone.
When two quantities are in direct proportion it means that:

- doubling one doubles the other
- trebling one trebles the other
- decreasing one by a quarter means the other decreases by a
  quarter
- and so on

The calculations are straightforward too – but, as always,
check that your answers make sense.

Here are a few examples, to explain how to do proportion
calculations. . . . . with checks to make sure that they <u>are</u>
proportion calculations.

1   You have a recipe for 6 people. This includes 800 grams of
    pasta. You need to feed 10 people. How much pasta will
    you need?
    [The number of people and the amount of pasta are in
    proportion – double one and you should double the other.]

2   The price of petrol goes up by 5%. You travel to work by
    car. Each week you used to spend £25 on petrol. How
    much will you spend on petrol now?
    [The price of petrol and the amount you spend on it are in
    proportion – double one and you double the other.]

3   The paint tin says a litre will cover 17 square metres. How
    much paint will you need to give two coats to a wall with an
    area of 27 square metres?
    [The amount of paint and the area it will cover are in
    proportion – double one and you double the other.]

Calculation Help desk

You have a recipe for 6 people. This includes 800 grams of pasta. You need to feed 10 people. How much pasta will you need?

If you are not sure how to work this out in one step,
first of all write down how to calculate how much pasta one person would need, then work out the amount for ten people.

6 people need 800 grams of pasta,
1 person needs 800 ÷ 6 grams of pasta
10 people need 800 ÷ 6 × 10 grams of pasta,
= 1333.3333 grams

Level of accuracy
This is far too exact an answer – working to the nearest 100 grams is probably quite accurate enough so about 1300 grams of pasta is a more sensible amount.

When two quantities are in inverse proportion it means that:
- doubling one quantity results in halving another
- trebling one results in the other being divided by three

Examples
If you double your speed, you halve your journey time.

If you treble the speed, your journey takes only a third as long.

Speed × 2 = journey time ÷ 2
Speed × 3 = journey time ÷ 3
Speed × 4 = journey time ÷ 4
Speed × 10 = journey time ÷ 10

Calculation Help desk

A journey normally takes 3 hours and a half and you average 50 miles per hour. On this occasion you leave quarter of an hour late. What do you have to increase your speed to, to arrive on time?

Journey takes

3.5 hours	distance = 50 × 3.5 miles	= 175 m
3.25 hours	speed = 50 × 3.5 ÷ 3.25 miles per hour	= 53.846 . . .mph.

Level of accuracy
It would be sufficient to say 54 miles per hour – or even 55 miles per hour, working to the nearest 5 miles per hour.

> **Remember** after working out the answer, you then need to check that the answer is to a sensible level of accuracy to make it realistic.

Using ratios

You may also come across proportion calculations, which include ratios. Sometimes the ratios are given in the form "1 to something", such as "1 to 50" or "1: 1000", such as in scale plans or maps. There is more about maps and plans in the next section.

You may also come across ratios like 3:4 or 1:3:6, where the ratios compare one quantity with another, or describe the proportions of different ingredients in a mixture.
For instance, when mixing concrete

Normal concrete mix 1:2:4 mix	Weaker concrete mix 1:3:6 mix
1 bucket of cement 2 buckets of sand 4 buckets of gravel Add water	1 bucket of cement 3 buckets of sand 6 buckets of gravel Add water

Sand and gravel are cheap to buy compared with cement, so if you don't need the full strength, the cheaper 1:3:6 mix may be adequate.

Calculation Help desk

You have to make 100 litres of dry mix 1:2:4 concrete.
How much cement, sand and gravel will you need?

For every 1 litre of cement, you need

         2 litres of sand

         4 litres of gravel

that is   7 litres altogether

For 1 litre of mix you need:

$\frac{1}{7}$ litre of cement,

$\frac{2}{7}$ of sand

$\frac{4}{7}$ litres of gravel

For 100 litres of mix you need:

$\frac{1}{7} \times 100 = 14.3$ litres of cement,

$\frac{2}{7} \times 100 = 28.6$ litres of sand

$\frac{4}{7} \times 100 = 57.1$ litres of gravel

Check your calculations by adding the three amounts. They should add up to 100 litres.

Level of accuracy: Working to the nearest litre is more than accurate enough – so you could quote the mixture as 14 l of cement, 29 l of sand and 57 l of gravel.

Scaling in drawings

Scale drawings and many maps show measurements 'to scale'. Once you know the scale of the drawing (such as 1 to 50, written 1: 50) you can measure distances on the drawing and multiply by the scale figure to find the full-size distance. In reverse (to convert from full size distances to distances on the drawing) you divide by the scale.

As well as multiplying or dividing by the scale figure, you may also have to change units between metres, centimetres and millimetres.

The drawing below shows the floor plan of an extension to a house.

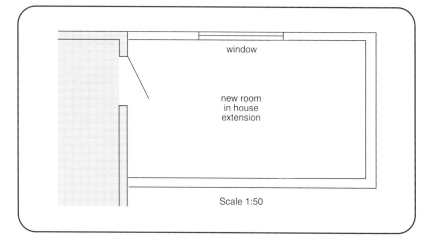

The scale of the drawing is 1:50, that is,

  1 millimetre on the drawing corresponds to 50 millimetres on the ground,

  1 centimetre on the drawing corresponds to 50 centimetres on the ground.

Calculation Help desk

1. The drawing shows the walls of the extension and where the door and window are.
   a) What is the length of the new room?
   b) What is the width of the new room?

a) The length of the room is 8 cm on the drawing
   The actual length is $8 \times 50 = 400$ cm = 4 m

b) The width of the room is 4.5 cm on the drawing
   The actual width is $4.5 \times 50 \div 100 = 2{,}25$ m

Compound measures

You will come across compound measures in all aspects of living – work, health, cooking, home, school, college, etc. Here are some examples of what are called 'compound measures':

Miles per hour, to measure speed

- 3 miles per hour is a steady walking speed
- 550 miles per hour is the speed of a jet airliner

Kilograms per square centimetre, to measure pressure

- 1 kg/cm^2 is normal air pressure
- 2.2 kg/cm^2 is the air pressure in many car tyres

Bytes per second to measure speed of transmission

- 204381 bytes at 42667 (bps) will take 4.7 secs to transmit

Milligrams per 100 millilitres, to measure concentration

- The driving limit is 80 mg/100 ml of alcohol in the blood

At 80 mg/100 ml how much alcohol is in the blood?

Most adults have about 5 litres – or 5000 millilitres – or 50 × 100 ml of blood in their systems.

80 mg/100 ml of alcohol means 80 × 50 = 4000 milligrams

(or 4 grams) of alcohol actually in their bloodstream

Note: above about 50 – 100 mg/100 ml alcohol in the blood, the individual's self control is reduced. Thinking and concentration are disturbed.

Milligrams per litre, to measure concentration

- 3.5 mg/litre is a typical amount of calcium in a bottled water

Deaths per 100 000, to compare causes of death

- about 420 per 100,000 males in the age group 45 – 74 die every year in the UK from heart disease
- about 300 per 100,000 males in the age group 45 – 74 die every year in the USA from heart disease

---

The / sign stands for 'per' or 'for every'.

Normal photocopy paper weighs 80 g/m^2

- 80 grams for every square metre

Better quality paper weighs 100 g/m^2

- 100 grams per square metre

Thin card weighs above 150 g/m^2

- 150 grams per square metre

Tonne-kilometres, to measure the scale of bulk transport being undertaken

- 200 million tonne-kilometres, means 200 million tonnes of coal being moved 1 kilometre
- 2 million tonnes being carried 100 kilometres, and so on.

Person-days, to measure how much time a job will take

- 40 person-days

– Here stands for ×.

The 40 person-day job could be done by

    20 people each working 2 days $20 \times 2 = 40$

    5 people each working 8 days $5 \times 8 = 40$

Formulas

(Note: sometimes we use **formulae**, the Latin plural)

Getting comfortable with formulas

You can learn how to use a calculator by using it to do simple calculations where you know the answers beforehand – write down the calculation – work it out on paper – then use your calculator to do it – the answers should be the same either way – so you can check you have used your calculator correctly.

In the same way you can get the hang of quite complicated formulas by checking in two ways.

Formulas are just shorthand ways of writing down 'recipes' for particular groups of calculations.

In a formula, you use letters to stand for values of what are called variables. Multiplication signs (x) between variables are often omitted in formulas.

Here is a familiar box shape (you may find it called a cuboid). The base of the box is shaded.

Below is the formula for its volume V, using l, b and h to stand for the length, breadth and height of the box.

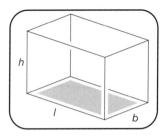

To work out the volume, you multiply length by breadth by height.

So the formula is
$$V = l \times b \times h$$
This is usually shortened to
$$V = lbh$$

Before using the formula you need to decide the units you are measuring in; this will then decide the units that the volume is in.

If l, b, and h are in **centimetres**, then V will be in **cubic centimetres** (usually written **cm³**).

If l, b, and h are in **metres**, then V will be in **cubic metres** (usually written **m³**).

The important point is that all the measurements are in the same 'family' of units.

You can make up formulas for yourself, too.

The base is a rectangle with l and b at right angles to each other.
The area of the base is is lb.

Starting from this, you can make up other formulas.

Write down a formula for the total area, A, of all the six sides of the box.
    [The formula is $A = lb + lb + lh + lh + bh + bh$],

It is equally correct if you have written the six 'terms' to the right of the = sign in a different order or if you have written hl instead of lh, for instance.

You can write the formula more concisely as $A = 2lb + 2lh + 2bh$

You may come across formulas that have the π (called 'pi') symbol in them.

'π' is the Greek letter for 'p'.

In calculations, π stands for the number you multiply the diameter of a circle by to find the length all the way around it (its perimeter or circumference). The Greeks were the first people to record the relationship between the circumference and the diameter of a circle, so it was natural to use π the Greek letter p.

Distance round a circle, its circumference, $c = \pi d$ where d is the diameter.

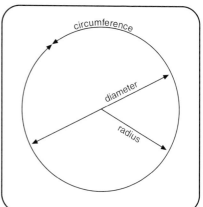

Another formula you may well meet is circle area, $A = \pi r^2$, where r is its radius.

π cannot be written as an exact fraction or decimal, but it is approximately 3.142.

You can use the π key on your calculator.

Often it is accurate enough to use 3.14 or just 3.1

To calculate the volume of a bucket, you can use the formula

$$V = \frac{\pi}{12}(D^2 + dD + d^2)h$$

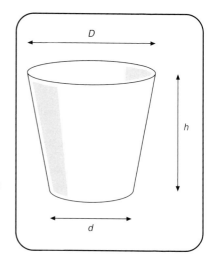

Find the volume of a bucket where $D = 20\,\text{cm}$, $d = 15\,\text{cm}$, $h = 21\,\text{cm}$

[Volume = 5086 cm³ = 5.086 litres ≈ 5.09 litres.]

Formulas and spreadsheets

You can set up formulas to make the computer spreadsheet carry out calculations for you.

This is one of the most powerful features of spreadsheets.

Each cell in a spreadsheet has its own reference, such as A1, B4, etc. – just like co-ordinates.

You use the = sign to start the formula.

You use * to mean x and / to mean ÷

The big advantage of using formulas in a spreadsheet is that if you change a number in the spreadsheet, any formulas you have included will change other numbers automatically.

This shows part of a spreadsheet for takings at a visitor centre shop one week

	A	B	C
1	Monday	£154.32	
2	Tuesday	£123.60	
3	Wednesday	£166.89	
4	Thursday	£190.31	
5	Friday	£203.20	
6	total for Mon – Fri	£838.32	
7	mean amount for 5 days		
8	total including VAT		

Here

cell A1 has Monday entered into it
cell B3 has '£166.89' entered into it
cell B4 has '£190.31' entered into it
cell B6 has '= B1 + B2 + B3 + B4 + B5' entered into it.

This means B6 = B1 + B2 + B3 + B4 + B5

The spreadsheet works out B1 + B2 + B3 + B4 + B5 and puts the result in cell B6

(i) What formula would you use to work out the mean (average) takings per week?
[B7 = B6/5]

(ii) What would you enter for cell B8 to make it work out the total of cell B6 with VAT of 17.5% added on?
[B8 = B6*1.175]

(iii) Experiment with a spreadsheet yourself. Invent your own formulas and try them out.

Rearranging formulas

A formula gives you a recipe for working out the value of one variable, the 'subject' of the formula, once you know the values of other variables. For instance once you know the measurements of a bucket, you can work out its volume.

A reliable way to do this is to
- put into words what the formula is saying
- see where that variable is which you want to make the new subject
- replace as many of the letters with any numbers you know
- reason backwards so as to end up with the subject on its own

Example

The formula for the volume of a cylindrical tank is

$$V = \pi r^2 h$$

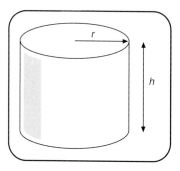

You want to find the height of a tank which has a radius of 0.75m and a volume of 5000 litres.

Working in metre units, you first need to change the litres into cubic metres.

There are 1000 litres in a cubic metre, so the tank will hold 5m³.

$$5 = \pi \times 0.75^2 \times h$$
$$5 = 176.7375 \times h$$
$$5 \div 1.767375 = h = 2.829\ldots m \approx 2.83\,m$$

Formula Help desk

Rearrange the cylinder volume formula to make r the subject

$V = \pi r^2 h$ Means volume = π times the square of radius, times the height of the cylinder.

$\dfrac{V}{\pi} = r^2 h$ Means volume divided by π = square of the radius times the height.

$\dfrac{V}{\pi h} = r^2$ Means volume divided by $\pi \times$ height = square of the radius.

If you divide by 2 and then by 3 the effect is the same as dividing by 6 Similarly, dividing by π and then by h is the same as dividing by $\pi \times h$ or πh.

We now have a formula for r^2.

$\sqrt{\left(\dfrac{V}{\pi h}\right)} = r$ To find r we must take the square root of $\dfrac{V}{\pi h}$

Find the radius of the cylinder, which is 1.6 m high and has a capacity of 3000 litres.

- Change the 3000 litres into 3 cubic metres
- Then you can do the calculation in two ways:
  - use the new formula or
  - put the numbers into the original formula.

Calculation Help desk

Using the new formula $r = \sqrt{\left(\dfrac{V}{\pi h}\right)}$,

so radius is $\sqrt{\left(\dfrac{3}{5}.0272\right)}$

$= 0.772 \ldots$ m
$= 0.73$ m to be sure of holding the 3000 litres

Using the original formula $V = \pi r^2 h$

So $3 = 3.142 \times r^2 \times 1.6$

So $3 = 5.0272 \times r^2$

So $\dfrac{3}{5}.0272 = r^2$

So $r = 0.73$ m as before.

Working out angles in right angled triangles

Triangle Help desk

- The sum of the three angles is always 180°
- The longest side is always opposite the largest angle
- The shortest side is always opposite the smallest angle

Consider triangle STU

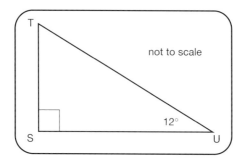

not to scale

You can work out the size of the angle at T (from S to T to U, or STU for short), in your head.

Angle TSU = 90°, angle TUS = 12°

Remember: The sum of the three angles in a triangle is always 180°

TSU + TUS = 90° + 12° = 102°

Angle STU = 180° − 102° = 78°
To check: 12° + 90° + 78° = 180°

Slopes and diagonals – right angled triangles

You may have to deal with situations involving slopes, angles and diagonals. Often these involve right angled triangles on their own, or rectangles divided in two by diagonals.

Right angled triangles

One of the oldest theorems in the study of triangles is due to the ancient Greek mathematician Pythagoras. It applies to right-angled triangles.

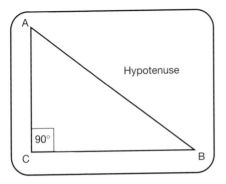

Definition
Hypotenuse = side opposite the right-angle
= longest side

Pythagoras's theorem states that, in any right-angled triangle, the square on the hypotenuse is equal to the sum of the squares on the other two sides.

$$AB^2 = BC^2 + AC^2$$

Example:

RS = 5 mm,
TS = 12 mm

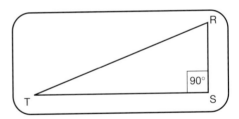

$RT^2 = RS^2 + TS^2$

$RT^2 = 5^2 + 12^2$

$RT^2 = 25 + 144$

$RT^2 = 169$

To find the length RT we must find the square root of 169

$RT = \sqrt{169}$

$RT = 13$ mm

Diagonals in rectangles and right angled triangles

There are situations when it can be useful to be able to work out the length of:

- The longest side in a right-angled triangle
- The diagonal of a rectangle

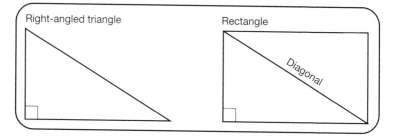

Right-angled triangle   Rectangle

Diagonal

Use the Theorem of Pythagoras to find the lengths.

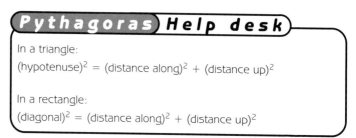

Pythagoras Help desk

In a triangle:

$(\text{hypotenuse})^2 = (\text{distance along})^2 + (\text{distance up})^2$

In a rectangle:

$(\text{diagonal})^2 = (\text{distance along})^2 + (\text{distance up})^2$

For instance you can use this to make sure that two lines (for instance the lines for two walls) are actually at right angles to each other, as in this example.

Here you are looking down on to walls KL and LM, which are at right angles to each other.

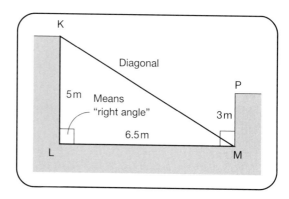

K

Diagonal

P

5 m   Means "right angle"

3 m

6.5 m

L   M

Questions

1. How long do you think the diagonal **KM** is – less than 6.5 metres, exactly 6.5 metres or more than 6.5 metres? [Imagine the line **KM** folding down on to **LM**, using the point **M** as a hinge. **KM** has got to be more than 6.5 metres long – maybe 7 or 8 metres or so]

   You can work out the exact length of KM using the formula above

Calculation Help desk

$KM^2 = KL^2 + LM^2$
$= 5^2 + 6.5^2$
$= 25 + 42.25$
$= 67.25$

so $KM = \sqrt{67.25} = 8.2006...$m
or 8201 mm to the nearest millimetre.

Try this yourself

2. How long should the distance KP be to the nearest millimetre? [KP = 6.801m]
   Hint: think of another right-angled triangle with KP as the diagonal.

You can work back from a known diagonal length to find another length.

3. A ramp board is 5 metres long.
   You use it to make a ramp for a rise of 700 millimetres.
   (i) How long a horizontal distance will it need?
   (ii) What will the gradient be?

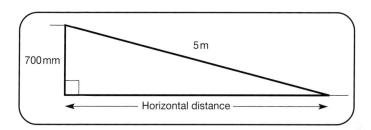

Calculation) H e l p   d e s k)

(i) Work in metres

(ramp length)^2 = (rise)^2 + (horizontal distance)^2

$5^2 = 0.7^2$ + (horizontal distance)^2

so $25 = 0.49$ + (horizontal distance)^2

so 24.51 = (horizontal distance)^2

$\sqrt{24.51}$ = horizontal distance

$4.951\,m$ = horizontal distance

Slopes, gradients and slope angles

Ramps improve access for wheelchair users, but they can have health and safety implications.

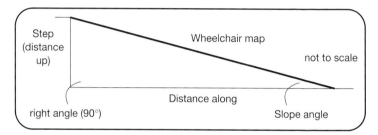

The diagram shows a wheelchair ramp.

The steeper the ramp (that is, the bigger the slope angle) the harder it will be to push a wheelchair up it – and the more dangerous it will be coming down.

Also, steep ramps can be hard to walk up and down on, especially for people wearing high heels or with stiff ankles. So it can be very important to work out the steepness of a ramp. Often the maximum slope is decided first, then you have to design the ramp accordingly.

You can describe the steepness of the slope

• by giving the slope as a gradient using:

  ○ a ratio   1 in 20 or 1:20

  ○ a fraction $\dfrac{1}{20}$ (meaning 1 unit up for 20 units along)

  ○ a decimal   0.05 $\left(\dfrac{1}{20} = 1 \div 20 = 0.05\right)$

• by giving the angle of the slope

These formulas are all you need to deal with situations involving gradients, slope and slope angles.

Formula Help desk

Tan of slope angle in degrees = $\dfrac{\text{distance up}}{\text{distance along}}$ = gradient

Example: $\qquad$ Tan $3° = \dfrac{1}{20} = 0.05$

or in reverse $\qquad$ slope angle in degrees = inverse tan of gradient

$$3° = \text{Tan}^{-1}\dfrac{1}{20}$$

Note: Tan is short for tangent

You can convert from gradient to slope angle (and slope angle back to gradient) using the 'tan' and 'tan-1' keys on your calculator.

Calculator Help desk

Trigonometric functions

Using the tan key	Work out tan 45°	
	Press the keys	Tan 45 =
	Or	45 Tan =
	Display shows	1
Using the tan^{-1} key	Work out tan^{-1} 1	
	Press the keys	shift tan 1 =
	Or	1 shift tan =
	Display shows	45

Further questions

4. For a 1 in 20 slope,
   (i) What is the distance along for a step up of 150 mm (15 cm)?
   (ii) What is the slope angle?

Calculation Help desk

(i) Gradient $= \dfrac{1}{20} = \dfrac{150}{\text{distance along}}$

distance along $= 150 \times 20 = 3000 \, \text{mm} = 3 \, \text{m}$

(ii) Change the gradient into a decimal

$1 \text{ in } 20 = \dfrac{1}{20} = 0.05$

Work out $\tan^{-1} 0.5 = 2.862. \ldots$
This is the slope angle in degrees.
To the nearest $0.1°$ this rounds to $2.9°$

5. Find the slope angles, to the nearest 0.1 degrees, for gradients of:
   (i) 1 in 25,   [2.3°]
   (ii) 1 in 40,   [1.4°]

Remember:
Press the = key to change the fraction to a decimal
Before you use the 'tan' key

Calculation Help desk

To calculate the gradient of a slope angle of $6°$ using a calculator

- Press 6 tan or tan 6 $0.105. \ldots$ (this is the gradient as a decimal.)

  It means the slope goes up 0.105 units for 1 unit along – that is $\dfrac{0.105}{1}$

- Press $\dfrac{1}{x}$ or x^{-1} $9.514 \ldots$ (this turns the fraction upside down.)

  So the gradient of 0.105 is $\dfrac{1}{9.51}$ or 1 in $9.51. \ldots$

To practice, start with your own choice of slope angle, use the 'tan' button on your calculator to find the gradient. To get back to the slope angle again use $\tan^{-1}$.

If you have done this correctly, you should come back to the angle you started with.

6. Find the gradient as a decimal and as "1 in something" for slope angles of
   (i) 2.5°   [0.04366. . . or approximately 1 in 23]
   (ii) 7 °   [0.1227. . . or approximately 1 in 8]
   Now to check, change them back to slope angles, using tan⁻¹

7. A ramp board is 5 metres long.

   You use it to make a ramp for a rise of 700 millimetres and a horizontal distance of 4.951 metres.

   What will the gradient be?

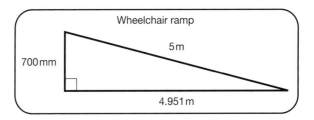

Wheelchair ramp
5 m
700 mm
4.951 m

Calculation Help desk

$$\text{Gradient} = \frac{\text{distance up}}{\text{distance along}} = \frac{0.7}{4.951} = 0.141$$

$\left(\text{using } \frac{1}{x} \text{ or } x^{-1} \text{ key}\right)$ about 1 in 7

Summary of results:

In a right-angled triangle:

$$d^2 = x^2 + y^2$$

$\frac{y}{x} = \text{tan of slope angle}$

$\tan^{-1} \frac{y}{x} = \text{slope angle}$

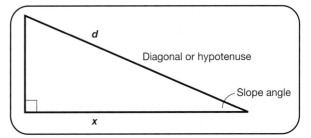

d
Diagonal or hypotenuse
Slope angle
x

x is the distance 'along'
y is the distance 'up'
d is the diagonal (sometimes called 'hypotenuse'), the longest (sloping) length in a right-angled triangle

What is trigonometry?

Trigonometry means 'measuring three cornered figures'. It shows the connection between the angles and the sides of a triangle. In particular, **trigonometric ratios** link two sides of a right-angled triangle with one of the smaller angles of the triangle.

The longest side of a right-angled triangle is the *hypotenuse* – this is the side opposite the right-angle.

The other sides are: the side **opposite** a given angle
the side **adjacent** (next to) a given angle

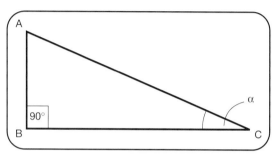

Trigonometric ratios

$$\text{Sin } \alpha = \frac{\text{side opposite}}{\text{Hypotenuse}} = \frac{AB}{AC}$$

$$\text{Cos } \alpha = \frac{\text{side adjacent}}{\text{Hypotenuse}} = \frac{BC}{AC}$$

$$\text{Tan } \alpha = \frac{\text{side opposite}}{\text{side adjacent}} = \frac{AB}{BC}$$

Calculator Help desk

sin, sin⁻¹, cos, cos⁻¹

Using sin	Work out sin 30	
	Press the keys	sin 30 =
	Display shows	0.5

Using sin⁻¹	Work out sin⁻¹ 0.5	
	Press the keys	shift sin 0.5 = or 0.5 shift sin =
	Display shows	30

Using cos	Work out Cos 60	
	Press the keys	Cos 60 =
	Display shows	0.5

Using cos⁻¹	Work out Cos-1 0.5	
	Press the keys	shift cos 0.5 = or 0.5 shift cos =
	Display shows	60

Example:

Use trigonometric ratios to find the missing sides and angle in triangle XYZ

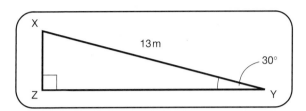

Calculation Help desk

i. Find the length ZY

$\text{Cos XYZ} = \dfrac{ZY}{13}$ $\text{Cos } 30° = \dfrac{ZY}{13}$

$13 \times \text{Cos } 30° = ZY$ $ZY = 11.26\,\text{m (to 2 d.p.)}$

ii. Find the length ZX

$\text{Sin XYZ} = \dfrac{ZX}{13}$ $\text{Sin } 30° = \dfrac{ZX}{13}$

$13 \times \text{Sin } 30° = ZX$ $ZX = 6.5\,\text{m}$

iii. Find angle ZXY

$\text{Tan ZXY} = \dfrac{ZY}{ZX}$ $\text{Tan ZXY} = \dfrac{11.26}{6.5} = 1.73$

$ZXY = \text{Tan}^{-1}(1.73) = 60°$

Memory Help desk

Try these to help you to remember the trigonometric ratios!

SOH CAH TOA

Two Old Angels, Sitting On High, Chatting About Heaven

Silly Old Harry, Caught A Herring, Trawling Off America

Six Old Horses, Clumsy And Heavy, Trod On Albert

$$S = \dfrac{O}{H}, \ C = \dfrac{A}{H}, \ T = \dfrac{O}{A}$$

Tables, graphs, charts, diagrams and statistics

Obtaining data

People usually obtain information (or 'data') for a purpose – to answer a question, to see if something is really true, to show that something that people believe is not true, to give them a better understanding of a situation. You can get hold of a lot of information by using the internet – for instance the UK national statistics office has a very useful website at www.statistics.gov.uk.

For instance, is it true that nearly everyone has access to a car so public transport is not really needed? The following table from national statistics provides interesting data.

Percentage of households having use of:						
	1981	**1994–5**	**1995–6**	**1996–7**	**1997–8**	**1998–9**
car	62	69	70	69	70	72
central heating	61	84	85	87	89	89
telephone	76	91	92	93	94	95
Source: ONC new earnings survey						

Question

Do you think nearly everyone nowadays has access to a car? [About three in 10 people live in households without a car, while in those with a car, often the car is used mainly by one person – so many people do not have access to a car. Contrast this with households with phones or central heating and see how these percentages have increased much more than the car percentages.]

A simple line graph shows the data more vividly.

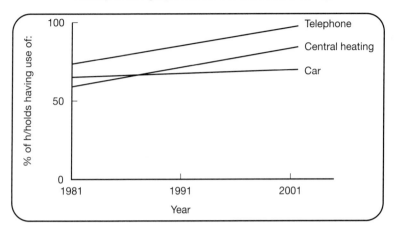

Understanding tables, charts and graphs

This section starts with a few 'warm-up' examples to get you looking at different graphs and charts.

The graph below enables you to convert petrol consumption in miles per gallon into litres per 100 kilometres.

38 miles per gallon converts to 7.4 litres per 100 km.

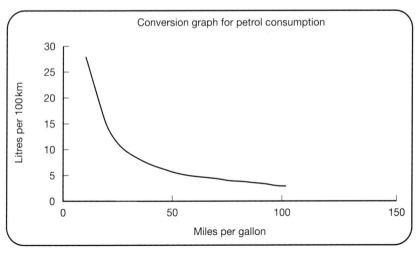

Question 1

Use the graph to change 10 litres/100 km into miles per gallon. [28.5 miles per gallon]

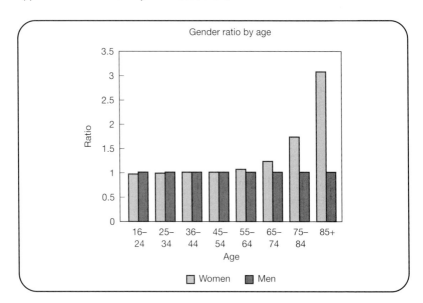

Gender ratio by age

Question 2
What comments can you make about this data?

Population by age in 1996 – Republic of Ireland			
Age group	**Total (thousands 000s)**	**Males (000s)**	**Females (000s)**
0–14	859	441	418
15–19	340	174	166
20–24	293	149	144
25–44	1016	503	513
45–54	412	209	203
55–59	154	78	76
60–64	138	69	69
65 and over	414	177	216
Total	3626	1800	1805
Source: Central Statistical Office Ireland			

Question:

> How do the ages of males and females compare in Ireland?
> [There are fewer females than males in 0–14 years, and for
> most of the blocks up to 50–59. Beyond the age-block
> 60–64, there are more females than males.]

Comparing Data

Using the range to compare

You can also compare sets of data using the range (highest value
– lowest value). However, often the range is not that helpful.
The table below shows the number of breakdowns per week for
two heavily used photocopiers over a three month period.

Breakdowns per week	Number of weeks for Machine A	Number of weeks for Machine B
0	3	1
1	5	2
2	2	4
3	1	5
4	1	1
5	1	

Using the mean to compare

In some cases it makes sense to compare two data sets using
their means. For instance in the last example, the total
number of breakdowns from each machine is important – and
you have to find this in calculating the mean number of
breakdowns per week.

Calculation Help desk

For machine A,

Total number of breakdowns	$= 21$
Total number of weeks	$= 13$
Mean number of breakdowns per week	$= \dfrac{21}{13} = 1.62$

Using the median to compare

Sometimes the median is a more reliable measure for comparison.

> ### Calculation Help desk
>
> For machine A,
> Total number of weeks = 13
> Median number of breakdowns
> will be in the 7th week = 1

Question:

(i) What is the mean number of breakdowns per week for machine B? $\left[\dfrac{29}{13} = 2.23\right]$

(ii) What is the median number of breakdowns for machine B? [2]

(iii) Which is the more reliable machine? [A]

	Machine A	Machine B
Range of breakdowns	5	4
Median number of breakdowns	1	
Mean number of breakdowns	1.62	

Working with grouped data

Often data is grouped before it reaches you. For instance, ages are often grouped in blocks of years, so you don't know how many people are 1 year old – or 2 years old, etc. All you know is the number of people in a given age group.

Comparing large data sets – using the median

You can find the median age by adding the total of all the rows to give the total population, and then finding which row half the total will be in.

The table on top of the next page shows the age make up for the Republic of Ireland in 1996.

Population by age in 1996 – Republic of Ireland			
Age group	Total (thousands 000s)	Males (000s)	Females (000s)
0–14	859	441	418
15–19	340	174	166
20–24	293	149	144
25–44	1016	503	513
45–54	412	209	203
55–59	154	78	76
60–64	138	69	69
65 and over	414	177	216
Total	3626	1800	1805
Source: Central Statistical Office Ireland			

Question
What was the median age in Ireland in 1996?

Calculation Help desk

A rough estimate

The median person will be $\dfrac{1\,813\,000 + 1\,814\,000}{2}$ = the $1\,813\,500^{\text{th}}$

This lies within the range 25 – 44

More accurate estimate:

The block '25 – 44' has $1\,016\,000$ people in it.

There are $1\,492\,000$ people aged below these ages, that is below 24.5,

$1\,813\,500 - 1\,492\,000 = 321\,500$

The median person will be $\dfrac{321\,500}{1\,016\,000^{\text{th}}}$ of the way up this block

The width of the block is 20 years so, assuming that people in the block are evenly spread out in age, the best estimate for the age of the median person is about

$(24.5 + \dfrac{321\,500}{1\,016\,000} \times 20)$ years

$24.5 + 6.3 = 30.8$ years

So the median age in Ireland is about 31 years.

Question

How does the age of the population of Ireland compare with those of Bangladesh and the UK?

- In Bangladesh the **median** age is 19.
- In the UK in 1988 the median age was about 30 for males and a little more for females.

Comparing medians gives a quick indication of how the ages of the populations compare.

[*The median age of the population of Ireland is similar to that for the UK. It is just over 10 years more than that for Bangladesh implying that life expectancy is lower in Bangladesh than in Ireland and the UK*]

Using the mean

Formula Help desk

To find the mean of grouped data:

$$\text{Mean} = \frac{\Sigma fx}{\Sigma f}$$

f = frequency

fx = frequency × middle value

Σ f = sum of all the frequencies

Σfx = sum of all (frequency × middle value)

How can you estimate the mean age of males in Ireland?

Calculation Help desk

- Assume that people are evenly spread out in each age block
- Assume that the middle value in each block is representative of the block

In the age group 0 –14 the middle value will be half-way between 0 and 14.5

$$14.5 \div 2 = 7.25$$

Then you multiply the middle value by the number (frequency) of males in that block.

$$7.25 \times 441 = 3197.25$$

Repeat this for each block.

Note that you have to choose a suitable middle value for the oldest block, say 74

Numbers in age groups for males in the Republic of Ireland for 1996			
Age group	**Males (000s) _f_**	**Middle age of each block (years) _x_**	**_fx_**
0 – 14	441	7.25	3197.25
15 – 19	174	17	2958
20 – 24	149	22	3278
25 – 44	503	34	17102
45 – 54	209	49	10241
55 – 59	78	57	4446
60 – 64	69	62	4278
65 and over	177	(74)	13098
Total	1800		58598.25
Source: Central Statistical Office Ireland			

Numbers in age groups for males in the Republic of Ireland for 1996

Calculation Help desk

$\Sigma fx = 58\,598\,250$

$\Sigma f = 1\,800\,000$

$\text{Mean} = \dfrac{\Sigma fx}{\Sigma f} = \dfrac{58\,598\,250}{1\,800\,000} = 32.6 \text{ years}$

Question:

(i) Work out the corresponding mean age for women.

   [$61\,003\,500 \div 1\,825\,000 = 33.4$ years]

(ii) Compare the means.

[_The mean age for females is less than one year higher than that for males_]

Presentation of data

A set of data will have little impact unless thought is given to its presentation. Tables, charts, graphs and diagrams can all be used to present data.

Tables are used to:
- Present original figures
- Summarise data
- Show distinct patterns in figures
- Provide information which may help to solve problems

Remember
- All tables must have a title
- The source of the data must be included (usually below the table)
- Columns and rows must have clear headings
- Units of measurement must be clearly shown

Regulations governing pre-school provision			
Age of child	**Staff:child ratio**	**Space per child**	**Lavatories/wash basins: child ratio**
Less than 2 years	1:3	3.7 m^2	1:10
2 to 3 years	1:4	2.8 m^2	1:10
Over 3 years	1:8	2.3 m^2	1:10
Extract: The Children Act 1989			

Charts and diagrams are used to:
- Give an immediate visual impression
- Make it easier to grasp the size of the figures involved
- Make it easier to make comparisons
- Make it easier to see how sets of figures are related

Remember

- All charts and diagrams must be as clear and simple as possible
- All charts and diagrams must serve an obvious purpose
- All charts and diagrams must have titles
- The source of the information should be indicated – usually at the bottom of the diagram
- All charts and diagrams should be clearly labelled
- All charts and diagrams must contain all the relevant information
- All charts and diagrams should be neat and attractive

Pie Charts

Pie charts are used:
- To show proportion
- To provide a strong, visual impact
- To provide an easy way to understand the relationship of the parts to the whole

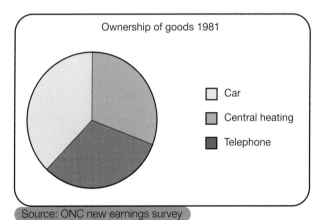

Ownership of goods 1981

Car
Central heating
Telephone

Source: ONC new earnings survey

Remember

- Pie charts are not very accurate
- Pie charts can be messy if too much information is included
- Angles must be calculated

Bar charts

Bar charts are used:

- To make comparisons
- Bar charts are easy to draw
- Bar charts are reasonably accurate
- Bar charts are easy to understand

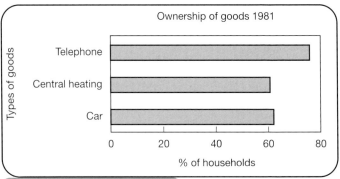

Source: ONC new earnings survey

Remember

- A bar chart has a vertical and a horizontal axis

- The axes must be labelled

- The scale of values is marked on the vertical axis

- The bars are of equal length and evenly spaced

- The height/length of the bar indicates the frequency

Compound (Multiple) bar charts are used:

- To make direct comparisons between two or more pieces of information
- To allow for quick comparison of several sets of information

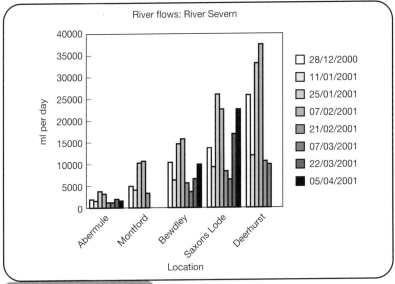

Source: Environment Agency

Component bar charts are used:

- To show how each part fits into the whole
- To allow us to compare one part with another part
- To allow comparison of proportion

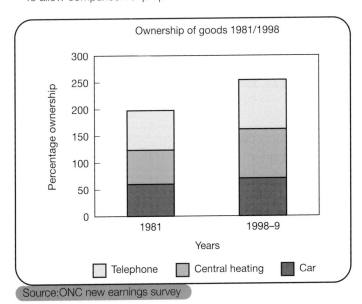

Source:ONC new earnings survey

Histograms are used to present:
- Grouped data
- Continuous data – i.e. data that is measured rather than counted

---

Remember

- A histogram consists of a series of touching bars or rectangles
- When data is grouped into even intervals the height of the bar indicates the frequency
- When data is grouped in unequal class intervals the area of a bar is proportional to the frequency

To calculate the height: Height (Frequency density) $= \dfrac{\text{frequency}}{\text{Class width}}$

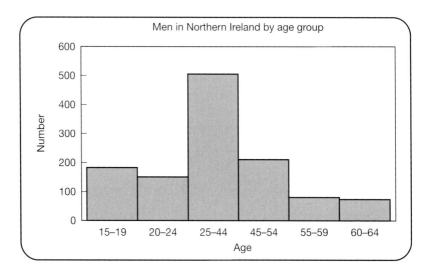

Men in Northern Ireland by age group

Graphs

Line graphs are used:
- To show changes over a period of time
- To make it easier to understand the figures
- To make it easier to compare by plotting several graphs on the same diagram
- Line graphs are the simplest graphs to draw

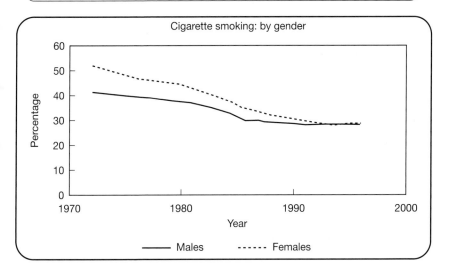

Cigarette smoking: by gender

Remember

- Time is ALWAYS shown on the horizontal axis
- Values are plotted using a 'x' and then joined by lines
- Can be complicated to read when there are several lines on the graph
- Scaling can be a problem when large numbers are involved
- Interpretation involves noting peaks and troughs and trying to explain them using a combination of knowledge and common sense

Scatter Graphs

Scatter graphs are used to show the relationship between two variables

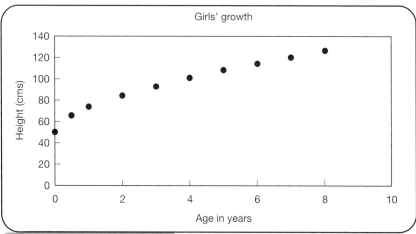

Source: Child Growth Foundation

Line of best fit

- A line of best fit is drawn on a scatter diagram where correlation is indicated
- It is a line that best fits through the points
- It can be used to interpolate and extrapolate values

Remember

- The line should be drawn so that the points are evenly distributed about the line
- It should follow the direction of the points
- The number of points lying above the line should be roughly equal to the number below
- It does not have to go through any of the points

10

Application of Number Level 3: Practising Skills in Part A

Numbers and Calculating

1 Use the x^y key on your calculator to find the values of
   (i) 2^{20} [1 048 576]
   (ii) 3^{12} [531 441]
   (iii) 5^{10} [9 765 625]

2 Without using your calculator, use the answers to the last question to work out the values of
   (i) 2^{21} [2 097 152]
   (ii) 3^{13} [1 594 323]
   (iii) 5^{11} [48 828 125]

3 Use your calculator to find the values of
   (i) 2^{-3} [0.125]
   (ii) 10^{-2} [0.01]
   (iii) 10^{-6} [0.000 001]

4 **Without using your calculator** work out the values of
   (i) 2^{19} [524 288]
   (ii) 3^{11} [177 147]
   (iii) 5^9 [1 953 125].

5   Try these for yourself:
   (i) $(6.2 + 1.45)^3$ [447.69712]
   (ii) $6.2 + 1.45^3$ [9.248625]
   (iii) $6.2^3 - 2 \times 1.45^3$ [232.23075

6   After how many years would £200 saved at 7% per year increase to at least £300?
> *[Work out £200 × 1.07=, continue pressing x1.07 = until you get over £300. You need to do this multiplication 6 times to reach £300. 15 Press 200 × 1.07 x^y 6 =. You will get £300. 14608 ≈ £300.15]*

Compare this with the example on page 144. Notice how the change in interest rate from 6% to 7% meant that your savings reached £300 a year earlier.

7   Use the x^y button on your calculator to find the totals after saving
- £300 over 6 years at 3.5%
- £300 over 6 years at 5.5%
- £500 over 7 years at 6.25%

[£368.78, £413.65, £764.31]

8   Experiment with your x^y button on your calculator to find the interest rate (to the nearest 0.1%) that would double the money you start with after saving for 8 years.
> *[After working out 1.07^8 (7% for 8 years), 1.08^8 and 1.09^8, if you try 1.091^8, you get 2.0072, so whatever you start with will have doubled if the rate is 9.1% or better]*

9   In a savings account, interest is added every month at a rate of 0.5%. The account is opened with a deposit of £250. How much is in the account at the end of 3 years?
> *[There are 36 months of interest to be added on, so the total amount at the end of the 3 month period is £250 × 1.005^{36} = £299.17]*

10 (i) Paul owes £900 to a loan company. The company adds interest of 2.5% per month on the amount he owes. How much does he owe after a year?
> *[Amount owed is £900 × 1.025^{12} = £1210.40]*

   (ii) What yearly interest rate is this equivalent to?
> *[His debt has increased by £310.40 so the annual rate is $\dfrac{310}{900} \times 100 ≈ 34.5\%$]*

11 Ann buys clothes from a catalogue. She buys a new outfit for £175. She pays back £12 per month until she has paid back the cost of the clothes including the interest on the amount she still has to repay. The catalogue firm charges 5% interest per month on loans.

How much does she owe after a year?

> [She starts by paying back £12 so she has £163 owing at the start.
>
> After 1 month this has grown to £171.15. She pays back £12 leaving £159.15 still to pay.
>
> After the 2nd month she owes £167.11. She pays back £12 leaving £155.11 still to pay.
>
> After the 3rd month she owes £162.87. She pays back £12, leaving £150.87 still to pay.
>
> After 12 months she still owes £108.30 even though she has paid back £144.
>
> [See how the debt doesn't go down by the same amount each month – each month it goes down by a little more than for the previous month]

Percentages and proportional change, ratios and scales

Increases and decreases

Try working out the results for these questions. In each case give your answers to a sensible level of accuracy

1   The price of petrol goes up by 5%. You travel to work by car. Each week you used to spend £25 on petrol. How much will you spend on petrol now?

2   The paint tin says a litre will cover 17 square metres. How much paint will you need to give two coats to a wall with an area of 27 square metres?

3   It takes you 20 minutes to cycle 3 miles. You are planning to cycle 10 miles. How long will it take?

> [1   The price has been increased by 5% – that is multiplied by 1.05, so the same will happen to the amount you spend per week. This is now $25 \times 1.05 =$ £26.25 – or a little over £26.]
>
> [2   1 square metre will need $1 \div 17$ litres, so 54 square metres will need $1 \div 17 \times 54$ litres = 3.176 litres. Again this is far too accurate; you can't be sure that one litre will cover exactly 17 square metres – it could be 16 or 18. Also you need to take into account that paint is often sold in 1 litre and $2\frac{1}{2}$ litre tins. A better answer might be 3.5 litres, using a 2.5 litre and a 1 litre tin.]
>
> [3   3 miles takes 20 minutes, so 1 mile could take $20 \div 3$ minutes, and 10 miles could take $20 \div 3 \times 10 =$ 66.6666 minutes, or better as "between 1 hour and an hour and a quarter allowing for slight changes in

speed". It might be safer to say up to an hour and a
half, to allow for winds, hills and you getting more
tired over a longer ride.]

4   A packet of fertiliser says "Enough for 25 square metres at
80 grams per square metre". You only need to spread it at a
rate of 50 grams per square metre. How much will it cover?

[At 1 gram per square metre it would cover
25 × 80 square metres, so at 50 grams per square
metre it will cover 25 × 80 ÷ 50 square metres =
40 square metres.]

Notice how in all these questions, after working out the
answer, you then need to check that the answer is to a
sensible level of accuracy to make it realistic.

Using ratios

In making marmalade, the ratio of fruit to sugar is about 1:2.

3 kilograms of marmalade	1 kilogram of marmalade
1 kilogram of oranges	$\frac{1}{3}$ kilogram of oranges
2 kilograms of sugar	$\frac{2}{3}$ kilogram of sugar

Of course, you add water and you boil the mixture so the
quantities are only approximate.

5   You plan to make enough marmalade to fill 12 jars, each
containing 340 g. What quantities of sugar and oranges
should you buy?

[12 of these jars would hold 4080 g or 4.08 kg – a little
more than 4 kg. So you will need about $\frac{4.08}{3}$ kg of
oranges, that is 1.36 kg of oranges, and double the
amount of sugar, that is 2.72 kg. A reasonable
shopping list would be 1.5 kg of marmalade oranges
and a 3 kg bag of sugar. This would give the correct
proportions of oranges to sugar, and allows for some
loss in quantity in boiling.]

Scaling in drawings

Centimetres and millimetres.

6   The drawing below shows the floor plan of an extension to
a house.

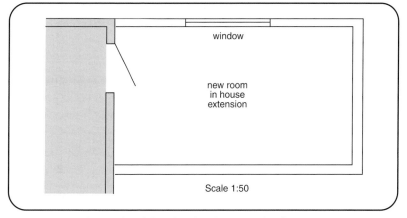

window

new room
in house
extension

Scale 1:50

The scale of the drawing is 1:50, that is,

1 millimetre on the drawing corresponds to 50 millimetres on the ground

1 centimetre on the drawing corresponds to 50 centimetres on the ground.

(i) The door opening should be 0.700 metres wide. Check the drawing to see if this is shown correctly.

On the drawing the door opening measures 16 mm which represents 16 × 50 = 800 mm or 0.800 m.
It should be shown as 14 mm which would represent 14 × 50 = 700 mm or 0.700 m

(ii) You can show the layout of furniture in the room. For instance, a sofa seen from above would be a rectangle. A sofa is 1.5 metres long by 70 cm from front to back. What should be the measurements of the rectangle to show the sofa to scale on the plan?

length on the plan is 1.5 ÷ 50 = 0.03 m or 3 cm
width on the plan is 0.70 ÷ 50 = 0.014 m or 1.4 cm

(iii) If the scale of the drawing was changed to 1:25 what would the length of the room on the drawing be?

The length of the room on the drawing would be 8 ÷ 25 = 0.32 m or 320 mm

7 The drawing below shows the plan of a field, drawn to a scale of 1:1000, with a footpath across it, between the two gates, and a house in the plot to the left. How close does the footpath come to the house?

[The closest the footpath comes to the house is 2.8 cm on the plan.]

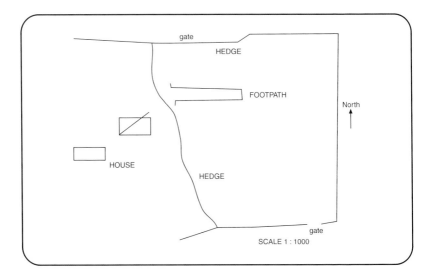

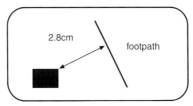

This represents 2.8 × 1000 = 2800 cm or 28 m. So the
footpath comes within about 28 metres of the house.
You can measure on the plan to the nearest 1 mm or 0.1 cm,
so the distance is accurate to no more than the nearest metre]

(i)   Roughly what is the area of the field?

*[The field is about the same area as a rectangle
65 m across by 70 m from top to bottom – so its
area is about 65 × 70 = 4550 m². You probably
cannot estimate the area of the field closer than to
the nearest 500 m². So the area is about 4500 m² or
0,45 hectare. A hectare = 10 000 square metres.]*

(ii) The footpath is moved to avoid crossing the field. Instead
of going straight across the field from gate to gate, it runs
from gate to gate, but now alongside the hedges on the
top (north) and the right (east) of the field, then back
along the bottom south hedge to the other gate. How
much longer is the new footpath than the old one?

> *[Length of old footpath on the plan is approximately 8.5 cm. New distance is roughly 14 cm, so extra distance is about 5.5 cm which represents 55 m]*

Compound measures

Now try these

8   There are 16 A4 sheets to the square metre. What is the weight of a 5 sheet A4 document in 80 g/m^2 paper?

> *[1 sheet weighs 80 ÷ 16 = 5 g, so the 5 sheet document will weigh 25 g – or more allowing for the extra weight of the print on it.]*

9   You want to pump up your car tyres to 32 psi (pounds per square inch). The tyre gauge shows pressure in kg/cm^2. What pressure in kg/cm^2 should you pump up the tyres to? (You may need to know that 1 kg = 2.2 pounds and 1 inch = 2.54 centimetres).

> *[You can work this out in steps, gradually converting from psi to kg/cm^2.*
> *32 psi = 32 ÷ 2.2 kg/square inch.*
> *1 square inch = 2.54 × 2.54 cm^2,*
> *so 32 psi = 32 ÷ 2.2 ÷ (2.54 × 2.54) kg/cm^2 = 2.25 kg/cm^2]*

10 Explain why, in the age group 45 – 74, the total male deaths from heart disease in the US are higher than in the UK, even though the death rate per 100 000 in the USA is lower. Use the figures above and assume that the population of the US is about 200 million compared to the population of the UK of about 60 million.

> *[There will be about 200 males in the age group 45–74 in the US for every 60 in the UK. This means that, for every 420 deaths in the UK in the age group 45–74 you would expect 300 × 200 ÷ 60 = 1000 per year in the same group in the US.]*

Formulas

A measure that is often used as a first check on a person's weight is the 'body mass index'. To calculate your BMI, divide your weight in kilograms by the square of your height in metres. If your BMI is between 20 and 25, then your weight is 'normal'. Note that if you train hard for sport, you may put on extra muscle weight, so your BMI may be above 25.

Use 1 kg = 2.2 pounds weight and 1 inch = 2.54 centimetres to convert from pounds and inches to kilograms and metres.

11 Write down a formula for BMI, using w kilograms as your weight and *h* metres as your height. Use it to find the BMI of a person weighing 75 kg who is 1.70 m high.

[$BMI = \dfrac{w}{h^2}$.

BMI of person is $\dfrac{75}{(1.70 \times 1.70)} = 25.95$, so this person is probably a little overweight]

If you wish, work out your own BMI.

12 A person wants to work out their minimum safe weight – that is their weight which would give a BMI of just 20. The person is 5 foot 5 inches high. What is their safe minimum weight in pounds?

[*A foot is 12 inches so the person is 65 inches high – that is about 165 cm or 1.65 m high.*

So $20 = \dfrac{w}{1.65^2}$,

that is $20 = \dfrac{w}{2.7225}$

This says that w divided by 2.7225 is 20, so w on its own will be 2.7225 times as much.

That is $w = 20 \times 2.7225$

$= 54.45$ *kg.*

This is 119.79 pounds or about 120 pounds (this is 8 stone 8 pounds using 14 pounds = 1 stone)]

13 For the same person as in 12, find their maximum safe weight.

[*Since BMI should be between 20 and 25, the maximum weight will be given by*

$w = 25 \times 2.7225$

$= 68.0625$ *kg*

$= 149.74$ *pounds or about 150 pounds*

Notice that you could have just used proportion to get this second answer directly.

You could just say the ratio of maximum to minimum weight is 25:20 so the maximum weight will be

$\dfrac{25}{20} \times 120$ *pounds* $= 150$ *pounds]*

If you wish, work out your own approximate maximum and minimum safe weights.

Measuring, Quantities and Space

1 What is the area of a circular paved area, with a diameter of 15 metres?

[*The radius is 7.5 metres so the area is* $3.142 \times 7.5 \times 7.5 = 176.73\,m^2$ *or about* $177\,m^2$]

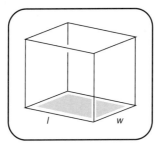

2   Write down a formula for the total area of the four vertical sides of the box. Call it A_v.

   [$A_v = 2lh + 2wh$]

3   What is the value of A_v if / is 18cm, w is 15 cm and h is 24 cm?

   [Av will be in square centimetres.

   $A_v = 2 \times 18 \times 24 + 2 \times 15 \times 24$, which is 1584 cm^2]

4   What is the area of a circular paved area, with a diameter of 15 metres?

   [The radius is 7.5 metres so the area is $3.142 \times 7.5 \times 7.5 = 176.73$ m^2 or about 177 m^2]

   To calculate the volume of a bucket you use the formula:

   $$V = \frac{\pi}{12}(D^2 + dD + d^2)h$$

5   Use the formula with measurements from a domestic bucket. Many buckets are labelled as '5 litre'.

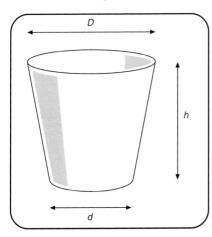

How accurate is this? Check the results of your calculations by measuring the volume of the bucket using water and a measuring jug.

Slopes and diagonals – right angled triangles

You may have to deal with situations involving slopes, angles and diagonals. Often these involve right angled triangles on their own, or rectangles divided in two by diagonals.

Slopes, gradients and slope angles

6   For the ramp below, the angle QPR (from Q to P to R) is 84°. What is the slope angle?

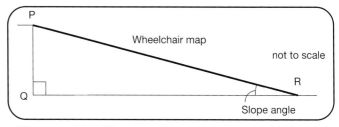

[Slope angle = 6°]

7 What is the minimum slope of a ramp for a step of 20 cm if there is 5 metres maximum for the distance along?
[20 up for 500 along = 1 to 25]

8 What is the gradient of a slope where the distance up is 500 mm and the distance along is 4 metres? (Hint: convert one of the measurements so that both distances are in the same units.)
[Working in millimetres, gradient is $\frac{500}{4000} = \frac{1}{8}$ or 1 in 8 or 0.125]

9 Find the slope angle to the nearest 0.1 degrees for the gradient 1 in 10

10 What is the slope angle for a gradient of 1 in 1? [45]

11 From what you have worked out already, estimate the decimal gradient for a slope angle of 6°
[It will be a little more than 0.1, as a slope of 0.1 gave a slope angle of 5.7°]

12 For the slope angles below, find the decimal gradients, and change them into "1 in something" form – all rounded to nearest whole number.
(i) 2.0°,   [0.0349. . . = 1 in 29]
(ii) 3°   [1 in 19]
(iii) 4°   [1 in 14]

13 Here you are looking down on to walls KL and LM which are at right angles to each other.

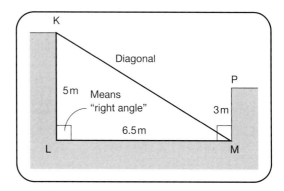

How long should the diagonal LP be, to the nearest millimetre?

[$LP^2 = 51.25$ so $LP = 7.159\,mm$]

14 Look at the wheelchair ramp.

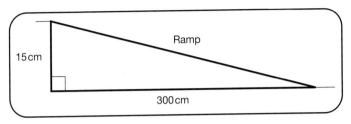

15 cm

Ramp

300 cm

What is the length of the ramp, to the nearest millimetre?

[working in centimetres, (ramp length)$^2 = 15^2 + 300^2$, so ramp length is 300.4 cm, or 3.004 metres. It is only slightly more than the horizontal distance of 3 m because the slope angle is small.]

15 In the diagram below, a rectangular metal frame has a diagonal to keep its corners at right-angles. What is the length of the diagonal?

[diagonal$^2 = 2.5^2 + 1.8^2$, so diagonal$^2 = 9.49$ so length of diagonal for corners of rectangle to be right angles is $\sqrt{9.49} = 3.081\,m$]

Diagonal

2.500 m

1.8 m

16 Find the gradient of a ramp board 5 m long used to make a ramp for a rise of 900 mm.

[The horizontal distance is 4.918, or about 3 centimetres less than before. The gradient is $\dfrac{0.9}{4.918} = 0.183$ or about 1 in 5.5]

17 The diagram below shows the rafters for a house roof extension. The rafters are 3.35 metres long, and their lower ends resting on the walls of the extension are 4.82 metres apart.

How high are the tops of the rafters above the walls, and at what angle from the horizontal are they sloping?

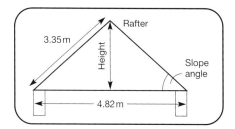

[*height*² + 2.41² = 3.35², so
*height*² = 11.2225 – 5.8081
= 5.4144 so height = 2.33 m
tan of slope angle = 0.9414.
. ., so slope angle = 44°*]

Understanding tables, charts and graphs

Obtaining Data

The graph below shows values of the so-called Body Mass Index for different heights and weights of adults. The two lines show upper and lower boundaries for the advised Body Mass Index for most adults. People with BMI below 20 are at risk of being underweight, whilst people with a BMI over 25 could be overweight.

Body Mass Index (BMI) Chart
Men and women

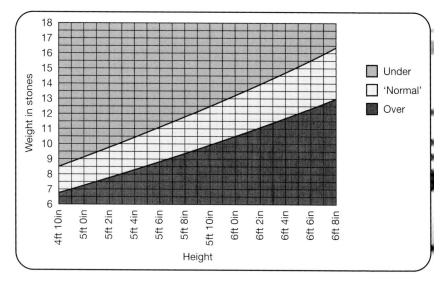

1   Use the graph to estimate the lower and upper bounds for advised weights for someone who is 5 foot 8 inches tall. You will need to convert this to metres first of all. Use 1 inch ≈ 2.5 cm and 1 foot = 30.5 cm.

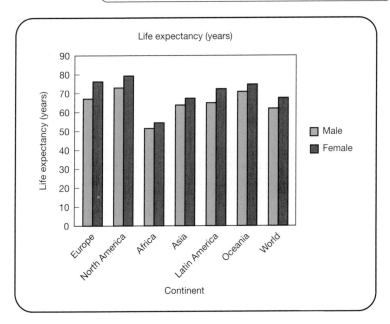

Continent	Males	Females
Europe	68	77
North America	74	80
Africa	52	55
Asia	65	68
Latin America & Caribbean	66	73
Oceania	72	76
World	63	68

Life Expectancy (years)
Source: United Nations Social Trends 1999

2 Convert these weights to kilograms using 0.45 Kg = 1lb (pound) and 1 stone = 14 pounds.

3 The table and bar chart overleaf are used to present data relating to life expectancy.
   (i) Which method of presentation is easier to understand?
   (ii) Comment on the data.

   [Life expectancy varies considerably throughout the world. People in the less developed regions have a much lower life expectancy than those in the developed regions. Africa has the lowest life expectancy of 52 years for males and 55 years for females. This compares with 74 and 80 respectively in North America, which is the continent with the longest life expectancy. In all continents the life expectancy of women is higher than that of men.]

4 Households with regular use of a car in Great Britain
   Percentages

Year	One car only	Two or more cars	Without a car	All
1961	29	2	69	100
1971	44	8	48	100
1981	45	15	40	100
1991	45	23	32	100
1997	45	25	30	100
Source: Department of the Environment, Transport and the Regions				

[The proportion of households with regular access to one car has remained stable since 1971 at around 45%. The proportion of households with access to two or more cars has increased considerably over the same period, from around 8% in 1971 to 25% of all households in 1997.]

5 Find the mean life expectancy for women

UK:Gender Ratio by age 1993

Age range	Women:Men
16 – 24	0.95
25 – 34	0.97
35 – 44	0.99
45 – 54	1.00
55 – 64	1.04
65 – 74	1.22
75 – 84	1.70
85 and over	3.06
All ages	**1.07**

Source: Office of Population Censuses and Surveys; General Register Office (Scotland); General register Office(Northern Ireland)

[Mean life expectancy $= \dfrac{711.8}{10.93} = 65.12$

Mean life expectancy for women = 65 years]

6 You are given two sets of data about smoking. Comment on the information.

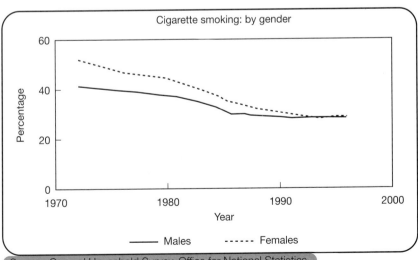

Cigarette smoking: by gender

Source: General Household Survey, Office for National Statistics

[Graph shows the fall in the proportion of men and women smoking cigarettes between 1972 and 1992–3. Twenty years ago men were much more likely than women to be smokers, but the proportions are now more or less equal at just under 30%]

Regular cigarette smoking among children aged 11 to 15 in England

	1982	**1986**	**1992**	**1996**
Boys	11	7	9	11
Girls	11	12	10	15

All figures are percentages.
Regular smokers are those smoking at least one cigarette a week.
Source: Smoking Among Secondary Children Survey, Office for National Statistics

[Smoking is increasing among children and among teenage girls in particular. In 1996, by the age of 15, a third of girls and over a quarter of boys were regular smokers.]

Presentation of data

1   In 1991 cooking appliances were the main sources of ignition for 25,000 accidental dwelling fires. Electric cookers accounted for 16,800 fires, gas for 7,200 and other cooking appliances for the remaining 1,000 fires.

Source: Gloucestershire Fire and Rescue Service Plan 1994–1995

Tabulate this data and produce a pie chart to illustrate the information.

In the following questions, you must decide how to present the data that is given. You must explain your decision.

2   How far do people move?

DISTANCE	1997	2001
5 miles or less	25%	2%
6 – 10 miles	30%	20%
11 – 50 miles	37%	63%
50 miles or more	8%	16%
Source: The Observer 16/9/01		

3   River Flows – River Severn

Mean daily flows Ml/day

	25/01/2001	07/02/2001	21/02/2001	07/03/2001	22/03/2001	05/04/2001
Abermule	3612	2925	990	970	1810	1700
Bewdley	14634	15780	5285	3520	6245	9815
Saxons Lode	26176	22865	8360	6560	16810	22650

Source: Environment Agency

4   Median statures of the adults of various nationalities

Nationality	Men	Women
British	1740	1610
US	1755	1625
French	1715	1600
German	1745	1636
Swedish	1740	1640
Swiss	1690	1590
Polish	1695	1575
Japanese	1655	1530
Hong Kong Chinese	1680	1555
Indian	1640	1515
Source: New Metric Handbook		

5 Hours spent watching TV

Hours	Nation	Men	Women
0 – 2 hours	45%	51%	39%
3 – 5 hours	42%	39%	45%
6+	12%	10%	14%
Source: Radio Times			

6 River flows – River Trent

	11/01/01	25/01/01	08/02/01	22/02/01	08/03/01	22/03/01	05/04/01
Drakelow	3710	11132	14140	3295	5570	6125	10755
Colwick	10655	23164	36080	8150	14315	11225	21940
Source: Environment Agency							

7 Times taken by bricklaying gangs to build 50 detached houses

Time in hours	Number of houses
240 – 249	1
250 – 259	1
260 – 269	7
270 – 279	10
280 – 289	16
290 – 299	11
300 – 309	2
310 – 319	2

8   Median statures of children from birth to 18 year

Ages	Boys	Girls
New – born infants	500	500
1 year old	715	715
2 years old	930	890
3 years old	990	970
4 years old	1050	1050
5 years old	1110	1100
6 years old	1170	1160
7 years old	1230	1220
8 years old	1290	1280
9 years old	1330	1330
10 years old	1390	1390
11 years old	1430	1440
12 years old	1490	1500
13 years old	1550	1550
14 years old	1630	1590
15 years old	1690	1610
16 years old	1730	1620
17 years old	1750	1620
18 years old	1760	1620

Source: New Metric Handbook

9   Population of Ireland 1901–1996

YEAR	TOTAL	MALES	FEMALES
1901	3 221 823	1 610 085	1 611 738
1911	3 139 688	1 589 509	1 550 179
1926	2 971 992	1 506 889	1 465 103
1936	2 968 420	1 520 454	1 447 966
1946	2 955 107	1 494 877	1 460 230
1951	2 960 593	1 506 597	1 453 996
1961	2 818 341	1 416 549	1 401 792
1971	2 978 248	1 495 760	1 482 488
1981	3 443 405	1 729 354	1 714 051
1991	3 525 719	1 753 418	1 772 301
1996	3 626 087	1 800 232	1 825 855

Source: Central Statistics Office, Ireland

10 Lengths of timber delivered to a building site

Length (cms)	Number of pieces of timber
60 – 62	5
63 – 65	18
66 – 68	42
69 – 71	29
72 – 74	6

11

Apply Your Skills at Level 3

Understanding Part B

Part B of the specifications is about the application of number
skills in the context of a specific activity, which may be part of
your main programme of study.

You will need to build a 'portfolio' of evidence showing how
you have met all the requirements of the Level 3 Application
of Number specifications.

- All components of Part B of the specifications must be covered.
- All the assessment criteria must be fully met

You must be able to plan and carry through at least one
substantial and **complex** activity in which you:
Obtain and interpret information

- Plan your approach to obtaining and using information
- Choose appropriate methods for obtaining the results you need
- Justify your choice

Use this information when carrying out calculations

- Carry out multi-stage calculations, including the use of a
  large data set and rearrangement of formulae

Explain how the results of calculations meet the purpose of
the activity

- Justify your choice of presentation methods
- Explain the results of your calculations

Definition Help desk

- **Substantial activity** – an activity that includes a number of related tasks where the results of one task are required to carry out subsequent tasks.
  For example, an activity involves obtaining and interpreting information, use this information when carrying out calculations and explaining how the results of the calculations meet the purpose of the activity
- **Complex activity** – an activity that can be broken down into a series of tasks and where the techniques needed to carry out the activity involve inter-related multi-stage calculations. The results obtained from individual tasks contribute to and help to determine the final results of your investigation

PORTFOLIO BUILDING

A portfolio is usually a file or folder that you use to present the evidence that shows how you have met the requirements for Part B of the Application of Number specifications. Portfolio building is very important. You will need to plan and organise your work from the start.

Start to build your portfolio as soon as possible. You can always remove work later if you find that you have produced something better.

Portfolio Help desk

Your portfolio should include:

Portfolio documentation that must:

- state which aspects of the unit you are claiming
- state exactly what work you have produced
- describe how this work covers the relevant parts of the unit
- state where the evidence can be found by the use of a clear referencing system
- be signed by an appropriate member of staff

Assignment briefs or tasks

Evidence of achievement showing:

- You have planned and carried through at least one substantial and complex activity that included tasks for N3.1, N3.2 and N3.3
- Shorter additional activities to cover any outstanding evidence (e.g. types of calculations, sources of information or forms of graphical presentation)

Why do I need portfolio documentation?

- To help you to organise your work
- To make it easy for a moderator to see that you have sufficient evidence
- To make it easy for a moderator to find the evidence
- To show a moderator where to find pieces of evidence that are not kept in your portfolio

What can I use as evidence?

- Evidence can be taken from a range of contexts (e.g. your programme of study, work experience, community activities and voluntary work)
- Evidence can be hand-produced or electronically produced and may include:
  ○ Written material, including calculations
  ○ Draft material showing the preliminary calculations etc.
  ○ Visual material such as graphs, charts and diagrams
  ○ 3-D scale models
  ○ Records of information that you have obtained e.g. surveys
  ○ Copies of source materials e.g. data from the internet

Quality is more important than quantity.

Definition Help desk

- **Moderation**
  The reassessment of a sample of portfolios to check that work is being assessed accurately and consistently to the National Standards.
- **Internal moderator**
  A key skills specialist appointed in your school/college/training agency. The internal moderator checks that your evidence has been assessed accurately and consistently to the National Standards.
- **External moderator**
  A person appointed by the Awarding Body. The external moderator checks that your evidence meets the agreed National Standards.

Providing the evidence

Evidence Help desk

You must:

- Identify a task which requires number skills
- Find the relevant numerical information needed to carry out the task
- Interpret the information correctly
- Make appropriate calculations accurately
- Present findings in written and graphical forms

Example:

A regular feature in the Sunday travel section of a national newspaper is a reader's request for suggestions of suitable holiday destinations. Certain constraints are given, e.g. cost, number in the party, type of holiday, time of year, etc.

The article in response to the request provides details of at least two alternative holiday destinations. It includes information on accommodation, costs, travel, etc

If you were a researcher on the paper how would you tackle such a task?

You must:

- Plan the activity
- Identify the tasks
- Carry out the tasks
- Present the results

PLAN, AND INTERPRET INFORMATION

- Decide on the make-up of the group. For example:
  ○ Friends
  ○ A family of 2 adults and 2 children
  ○ Party of students
  ○ Business group
  ○ Senior citizens
- Set a financial limit:
  ○ Luxury holiday
  ○ Family holiday
  ○ Budget holiday
- Investigate destinations:
  ○ Travel agents
  ○ Internet
  ○ Specialist companies

- Time of year:
  - Spring
  - Summer sun
  - Autumn
  - Winter snow
- Type of holiday:
  - Activity holiday
  - Beach holiday
  - Ski-ing holiday
  - Specialist interest holiday
  - Business
- Type of accommodation:
  - Camping
  - Hostel
  - Hotel
  - Self-catering
  - Cruise ship
  - Chalet

PLAN, AND INTERPRET INFORMATION

Evidence:
- Clear description of the activity
- The purpose of the activity
- Details of how you expect to obtain the relevant information from **2 different types** of sources
- Clear sequence of tasks showing how you intend to use this information
- Target dates for carrying out the tasks
- Evidence that you can obtain information
- Justification of your choice of the mathematical methods that you intend to use in your calculations

---

CARRY OUT CALCULATIONS

- Costing calculations
  - Cost of travel
  - Cost of accommodation
  - Supplements/reductions
  - Extras e.g. trips
  - Car hire

- Currency exchange rates
  - Could provide large data sets for comparison using statistical measures
  - Conversion calculations
- Travel arrangements
  - Passports/visas
  - Travel to airport
  - Cost of travel to airport
  - Time at airport
  - Length of flight
  - Travel from the airport
- Weather could provide large data sets for comparison using statistical measures
  - Temperatures
  - Sunshine
  - Rainfall/snowfall
- Clothing
  - Beachwear
  - Ski-wear
  - Hiking gear
  - Wet weather clothing
  - Specialist clothing
- Special considerations
  - Vaccinations
  - Preventative medicines

CARRY OUT CALCULATIONS

Evidence:

- Most calculations should involve at least 2 stages
- There must be at least one example from each of the 4 categories
- Clear demonstration of the methods used
- Indication and justification of the levels of accuracy used
- Details of how you have checked methods and results
- Indication of where corrections were needed

INTERPRET RESULTS AND PRESENT FINDINGS

- Decide how you will present your findings
  - Briefing
  - Report
  - Article
  - ....

- Detailed costing calculations
- Detailed travel arrangements and timings
- Recommendations

Examples of graphs, charts and diagrams

- Graphs
  - Exchange rates
  - Temperatures
  - Rainfall/snowfall
  - Hours of sunshine
  - Currency conversion
- Charts
  - Sunshine
  - Rainfall/snowfall
  - Temperature
  - Currency conversion
- Diagrams
  - Scale drawing of accommodation
  - Map of area
  - Flowchart for travel schedule

Evidence Help desk

INTERPRET RESULTS AND PRESENT FINDINGS

Evidence:

- Descriptions of the methods used to obtain your results
- Explanations of the results of calculations in terms of how they met the purpose of the activity
- Selection and justification of the choice of methods of presentation
- Effective presentation of findings using forms that are appropriate to the nature of the data being presented
- Use of at least one graph, one chart and one diagram
- When IT is used to produce graphs, charts and diagrams their accuracy must be checked and they must be explained fully
  The presentation must be of 'publishable quality'

Appendix: Sample Action Plans

Application of Number Action Plan

The skills I need to learn or improve	What I need to do to learn or improve the skills	By when	Tutor's comments

Application of Number Action Plan

The skills I need to learn or improve	What I need to do to learn or improve the skills	By when	Tutor's comments

Index

page numbers in italic refer to charts and illustrations
L1 = Level 1; L2 = Level 2; L3 = Level 3